Praise for *Mending the Mind*

'With startling openness and honesty, but without angling for sympathy, Kamm describes his symptoms . . . Kamm's overall message is that despite the utter grimness of those symptoms, sufferers should remain optimistic . . . *Mending the Mind* reminds us that, despite our hazy understanding of depression, and despite the true horror of the illness, some hope for recovery remains' *The Times*

'The combination of Kamm's up close and personal and investigative approaches together make this both useful and insightful. It's also extremely intelligent, compassionate and well-written' *ndard*

'Uplifting . . . easy-to- of the illness, such as what ca logical treatment, living with *Times*

'A tour de force that is not just personal, but looks at depression through science, art, literature and history. The combination makes it an important, affecting and effective book'

Alastair Campbell

'Oliver Kamm's urbanity, erudition and compassion are raised to the power of two in *Mending the Mind*. He put them to work in crafting this gorgeous and urgent book, and on every page they remind us of his moral that enviable gifts are no protection against the affliction of depression'

Steven Pinker, Johnstone Professor of Psychology, Harvard University, and author of *How the Mind Works*

'There are lots of books on depression, but not many combine both an authentic and moving personal story with rigorous research and analysis. Kamm tackles the big questions on depression, and his answers are clear, unsentimental and compelling'

Professor Sir Simon Wessely, Regius Chair of Psychiatry, King's College London

'*Mending the Mind* elegantly combines a powerful personal story of experiencing and recovering from severe depressive illness with a wide-ranging overview of what depression means to artists and scientists, poets and practitioners. Oliver Kamm has succeeded in putting his lived experience here and now at the heart of an inclusive, broad-minded account of how depression has been understood and misunderstood, diagnosed and treated, over hundreds of years'

Professor Edward Bullmore, Cambridge University, author of *The Inflamed Mind*

'This is a deeply personal and important book. It is painful but ultimately exhilarating to read. As Oliver Kamm notes, depression may be common but it is still not widely understood. At a time when isolation risks a heavy psychological toll, it deserves to be read more than ever'

Andrew Bailey, Governor of the Bank of England

Oliver Kamm is a leader writer and columnist for *The Times*. His book *Accidence Will Happen: The Non-Pedantic Guide to English* was published by Weidenfeld & Nicolson in 2015.

MENDING THE MIND

The Art and Science of Overcoming Clinical Depression

OLIVER KAMM

First published in Great Britain in 2021 by Weidenfeld & Nicolson
This paperback edition published in 2022 by Weidenfeld & Nicolson,
an imprint of The Orion Publishing Group Ltd
Carmelite House, 50 Victoria Embankment
London EC4Y 0DZ

An Hachette UK Company

1 3 5 7 9 10 8 6 4 2

A CIP catalogue record for this book is available from the British Library.

ISBN (Mass Market Paperback) 978 1 4746 1084 1
ISBN (eBook) 978 1 4746 1085 8
ISBN (Audio) 978 1 4091 8657 1

Typeset by Input Data Services Ltd, Somerset

Printed and bound in Great Britain by Clays Ltd, Elcograf S.p.A.

www.weidenfeldandnicolson.co.uk
www.orionbooks.co.uk

To the memory of my mother, Anthea Bell

Contents

Contents

[illegible]

PART 1 UNDERSTANDING DEPRESSION

PART 2 TREATING DEPRESSION

Preface

Clinical depression is weird. Nothing in life has puzzled me more. I bore no outward sign of injury, yet every waking moment was a torment that overwhelmed my thoughts and senses. All I could think of, on and on, was that the world I'd known was gone. Only agonies awaited me for the rest of time. I yearned for that time, for my time, to end. The condition lasted in a severe state for a year, stretching from one December to the next, and persisted at a lower level for a while longer.

Depression is classified in the medical manuals as a mental disorder. Once my illness had been diagnosed and I knew what it was, that noun phrase to me felt apt. Order was gone. The fixed points in life, an intuitive understanding of where I belonged and what my ambitions and purposes amounted to, had unravelled.

What took their place was an all-enveloping, suffocating anguish. While consumed by guilt and stricken with remorse, I retained sufficient awareness of my surroundings and vestigial rationality to sense that a mental condition in which nothing but the self matters is solipsistic. It's not creditable. It's undignified. It's a betrayal of those who trust in your judgement and depend on your care. The realisation intensified the shame and sense of delinquency that had brought me to this pass in the first place, and thus the downward spiral of the psyche continued.

In this book I describe what it's like to be mentally ill. I set out what's known in the literature, from scientific inquiry and first-hand accounts, of the nature of depression and how it differs from the low moods that everyone experiences. That distinction is crucial to understanding depression. The sadness of life is inescapable. For most people in modern western societies sadness will be episodic, and it will alternate or coexist with happier times. Depression, on the other hand, is radically unfamiliar. It unhinges us from everything we thought we knew about the world and estranges us from every other person. It makes us, in turn, strangers to those we love.

I further recount the stages and techniques employed in the necessarily lengthy process of getting out of that state. Because reliable medical knowledge of this disorder is limited, it used to be common to talk of a sufferer's depression as being in remission rather than being cured. The language of mental health has shifted in the last twenty years. My experience with effective therapies, which are founded on research and evidence, taught me how to recognise depressive thoughts as they emerged and to repel them before they dragged me down once more.

It's the most direct experience I've had of how pure theory can affect hard fact and alter it. The American psychiatrist Aaron Beck, a pioneer in this form of psychological treatment, has written: 'When a person is able to fill in the gap between an activating event [for mental disorder] and the emotional consequences, the puzzling reaction becomes understandable. With training, people are able to catch the rapid thoughts or images that occur between an event and the emotional response.'[1]

I can't say with certainty, but do believe with conviction, that – having received the requisite training – I made more than a full recovery from severe depression. My life became richer for knowing how, and for applying the methods I'd been taught. I would never have wished to go through clinical depression, but the process of recovery made me more at ease in a bleak and impersonal universe whose only meaning is what we feeble humans attribute to it.

The mind is mendable

I endeavour in this book to recount not only what it's like to have depression but also what I learnt about the condition while going through it and what I've found out about it since. I wanted to know what had happened to me and, having reflected upon those findings, then to tell others in the same state that their condition is not hopeless. Those inquiries have been inevitably partial and my account is fragmentary. I don't have the specialist knowledge or training to write anything resembling a comprehensive study of depression and the discoveries made about the mind by modern science. Instead, I offer a personal account of what happens when a mind malfunctions and shuts down, and how evidence-based treatments can help. The raw material of this book is drawn from experience and from the literature, in science and the humanities, of the dramatically altered perceptions characteristic of the person suffering depressive disorder. My conclusion will look prosaic on the page yet is urgent for human welfare: depression is real and it's mendable.

I resolved to set out the evidence for this observation because I believe it matters for a public debate on depression that isn't always as well informed, or as sensitively and sensibly conducted, as it might be. As well as wishing to interest general readers, I aim to reach those sceptics who cavil at the notion that clinical depression is a feature of modern society. Most of all, I hope to inform and reassure the many people whose lives are blighted by depression and who wonder if they will ever emerge from it.

If you suffer from depression or are close to someone who does, be assured there is a route out of it, and there are medical and other specialists who can help you find it and guide you along it. There are remedies that have been shown by clinical trials to work. They worked for me, and hard data shows that these psychological and pharmacological treatments will be of benefit for most patients with depression.

Depression in a time of crisis

I write this during the greatest peacetime public health crisis for more than fifty years, and by some estimates since the influenza pandemic of 1918–20. The global pandemic of 2020, of a novel virus originating in Wuhan, has infected millions and killed hundreds of thousands. While the coronavirus spreads a disease that attacks the human respiratory system, it also indirectly has an impact on people's mental health, owing to the strictly necessary remedial measures adopted by governments to halt the spread of infection. The most immediate effect is likely to be loneliness. Others include grief, trauma from domestic violence or family loss, and alcohol and substance abuse. And there are also possible direct effects on mental health. Covid-19 is a multi-system illness that has a neurological impact (there have been reports of seizures, confusion and cognitive impairment among its victims) and it may increase the incidence of psychiatric illnesses, as happened in the influenza pandemic of a century ago. Not all mental-health problems have effective treatments, and even for those that do, there may be difficulties in delivering them and in treating patients who have comorbidity. The global pandemic may decisively alter the debate on mental-health policy, and it's vital we get it right. Dr Elke Van Hoof, a Belgian clinical psychologist, has noted that around the world some 2.6 billion people are in some kind of lockdown, which is 'arguably the largest psychological experiment ever'.[2]

Few find it easy to be alone. Quarantine and lockdown during the crisis have ensured that many are. Critics of these measures have stressed the damage to the economy but also the psychological toll of keeping people isolated. There is, further, the impact on access to the rest of healthcare, such as cancer treatment, to take into account.

There's admittedly no disputing that these remedial measures have caused a huge economic downturn but it does not follow, as is claimed by critics (who include President Trump), that they will thereby cost more lives than they save. Perhaps counter-intuitively, economic research by the Nobel laureate Angus Deaton suggests that the short-term impact of recessions on life expectancy is typically to increase it (there are, for example, fewer fatal accidents because of less traffic on the roads). Deaths by suicide do increase, but these are a very small component of deaths from all causes.

Suicide is an urgent humanitarian issue regardless of the coronavirus crisis. Depression is the main contributor to those deaths. The crisis has undoubtedly wreaked emotional havoc among those worried for the future, isolated from loved ones or suffering the grief of bereavement. They need support. Whether their emotional states mutate into depression in a clinical sense may depend on it.

Everyone feels low and sad at times. It's part of the human condition. Low mood is what you'd expect when we can't visit friends or loved ones, engage in normal recreation, or even just change the scenery by getting on a bicycle or a train. It will particularly affect those who live in cramped or substandard accommodation, lack access to green spaces, and are in dysfunctional or abusive domestic relationships.

Depression is an illness not only of sadness, however, but of distorted thinking. It will be there in times of normality and placidity, as well as turmoil. And when we're left isolated, these distortions of thinking can be magnified by the constant process of rumination. Public Health England (PHE) issued a set of guidelines in the early stages of the spread of the coronavirus on the mental and wellbeing aspects of the crisis, which identify the threat. The document stresses: 'It is okay to acknowledge some things that are outside of your control right now but constant repetitive thoughts about the situation which lead you to feel anxious or overwhelmed are not helpful.'[3]

That is a compassionate message and an informed one. The emotional difficulties experienced by those in lockdown are not inevitably the same as illness. To prevent an epidemic of depressive disorder in those who are distressed by these measures and the death of loved ones places an onus on the rest of us. The means of staunching a spread of mental disorder are in our hands. Following the advice of PHE to stay in touch with people, to support others, to look after your physical wellbeing and to take time to focus on the present will make our society as well as ourselves more resilient.

Yet it will do no good, and may inflict appreciable harm, if the public assimilates from the crisis that, so far as mental health is concerned, all that's necessary is that we should learn habits of fortitude. Depression and its associated disorders are not a result of weakness or sloth. Nor are they, as some critics argue, a pathology of modern society. They are illnesses. You cannot wish yourself healthy in mind, any more than you can mend a broken limb by the power of prayer. But there are effective treatments, founded on evidence and critical inquiry. These don't require an exhaustive search into the personal psyche, or sweeping social change, to work. Even supposing these outcomes were valuable, we don't have the time to wait upon them – and we don't need to wait to help ease and dispel mental illness.

Testing and validating treatments is crucial. For those feeling low and worried, talking to a sympathetic outsider may prove invaluable in grounding them in a wider social network. For those suffering from clinical depression, it may not work at all or even make things worse, if the approach is wrong. That's what happened to me. There was no shortage of goodwill from those offering what are known as talking therapies, to which I was referred. But there was a lack of scepticism of dogma and procedure, of the type that characterises the scientific approach. Fortunately, I learnt that better ways were available. I craved them, sought them, benefited enduringly from them, and eventually found a way out.

Outline of the case

This book recounts what I learnt from the treatment, the science, the depiction of depression in poetry and art, and simple human companionship of a type I had known little of before. In the first part, 'Understanding Depression', I set out what we know about clinical depression and how it's been perceived historically. Chapter 1, 'What Is Depression?', delineates the characteristics of this most peculiar illness and recounts how I came to realise that this was what I had. It states my thesis that depression, in its mild or severe state, is in principle comprehensible and that it's capable of being cured.

In Chapter 2, 'What Causes Depression?', I survey the explanations that have been arrived at through the ages for a condition that seems to have afflicted humanity from prehistory onwards. My argument is that depression is mysterious but that it yields to scientific inquiry. We know that thought, cognition, memory and emotion are all produced by brain states. This understanding of the mind can provide us with clues to why depression occurs.

Chapter 3, 'How We Understand Depression', introduces the notion that the scientific assessment of depressive disorder needs to be supplemented with the accounts of writers and philosophers. We can't directly observe mental illness, but know it only through the symptoms of those who suffer it. The testimony and observations of poets, artists and thinkers are invaluable in expanding our knowledge of and speculation about depression.

In Chapter 4, 'How We Misunderstand Depression', I conversely describe how some stubborn misconceptions about mental illness permeate public debate. They've been prominent in the coronavirus pandemic. It's a very common belief that depression is a fashionable malady and that it can be remedied by teaching habits of mental resilience and adopting a bit of gumption. The ethos of the stiff upper lip is with us still, now garbed in critiques of a supposed 'therapy culture'. These are corrosive notions that do much avoidable harm and reinforce the stigma attaching to mental illness.

In the second part of the book, 'Treating Depression', the argument advances from the nature of the condition to the ways, historical and current, that it's been diagnosed and dealt with. Chapter 5, 'Diagnosing Depression', recounts my own experience of having the illness identified and explained to me. It was my first conscious encounter with mental disorder as opposed to sadness, and it came upon me as a complete surprise. There will be many sufferers who, like me, never consider the possibility that they are clinically ill rather than merely stressed and distressed. It's an important realisation that your mind is malfunctioning; to get well requires that you face recognisable problems that others have experienced too.

In Chapter 6, 'Medical Treatment', I give a synopsis of how physicians have over centuries sought to devise remedies for mental illness. Some of these are recognisably cruel and have fallen into disfavour; others are controversial because empirical evidence is scant, though they are still in use for extreme cases. But medical science has devised methods that help stabilise the wild swings of emotion characteristic of depressive disorder. There are numerous scare stories about, and some extravagant praise for, the class of pills known collectively as antidepressants. I explain what antidepressants are and why their benefits outweigh their risks.

In Chapter 7, 'Psychological Treatment', I turn to the often-contentious issue of talking therapies. The image of a patient reclining on a couch and giving free-form expression to their distant memories is the stuff of countless comedy sketches. Modern therapy, thankfully, isn't really like that;

for practical reasons (notably the time and expense involved) it's largely moved on from the theories expounded, though with only tangential relevance to depression, by the psychoanalysts who once held sway in the treatment of mental illness. I describe the different forms of talking therapy, in which the residue of the work of Freud and his followers still lingers.

Chapter 8, 'Living With Depression', describes what the illness was like for me and what it took to get me through it and out of it. The experience impressed upon me that there are human needs we can count as intrinsic. The most essential, which I'd never fully appreciated till I became mentally disturbed, is to know we are social animals. Our insignificance in vast expanses of time and space is ultimately made bearable by our knowledge that others share life's journey with us. This is true of those we know and love, and also of those who are chronologically distant yet whose experiences and perceptions we know vicariously through their work.

In Chapter 9, 'Depression and Art', I draw on the writings and imagery of those who've suffered depression and illuminated the nature of the disordered mind. Their ability to describe the experience and to depict the world in distinctive ways enabled me to shift the stubborn habits of thought that had brought me to despair. And it helped acclimatise me to a mental world that I had thought lost. In the Epilogue I describe the chasm between the disordered mind and the healthy one, the relief I felt when returning to sanity and my bafflement on having ever left it.

Becoming well

My experience of depression was unique to me, in the sense both of my never having had it previously and of its character, but the events that prompted it were unextraordinary and pedestrian. My life has been fortunate, with professional and personal satisfactions, and it would be hard to point to a defining event in it that would bring anyone low.

Disappointments, thwarted ambitions, rejections, failures and the pains of loneliness are the stuff of experience. In my circle of friends are many whose tribulations might have broken lesser people. They've witnessed and survived the genocide of their communities. They've experienced scarcely imaginable suffering and bereavement and withstood their adversity. There is no comparison between experiences like these and the fate of those of us who've lived all our lives in constitutional societies.

But depression isn't proportional to human woe. It has a very long history, and studying history helps us better understand the present. We know from the testimony of sufferers and observers that the same symptoms of severe depression occur in every age, though the condition has been categorised under many labels. Not then understanding depression or knowing how to cope with it, I determined to find out about it. For the first time, I understood depression no longer as a speculative and ill-defined term to describe a range of mental afflictions that happen to others, but as a specific and identifiable state that had befallen me. I learnt what I ought already to have known: depression is an enduring human condition that creates havoc and misery. And I resolved to write about it one day in the hope of explaining this crushing affliction to those who haven't known it, and to help others who are experiencing it now.

I'm anxious that more people should know that depression can be countered and overcome by therapeutic and medical interventions, principally the so-called talking therapies along with antidepressants. I'm an advocate of certain of these treatments, not only because they foreshorten needless suffering but because recovery is only a limited aim. The mind can be mended to guard against the experience of depression happening again, and to speed its departure if it does reoccur. Mental-health practitioners increasingly speak of making sufferers from depression well, rather than better. Some even maintain that the patient can become 'better than well'.

That final claim may appear to have an ominous hint of social engineering to it (though, as we'll see, it's more about chemical engineering, using medication to alter the chemical balance within the brain). I dislike the phrase for that reason but, at least with regard to psychological techniques rather than drugs, it's got more than a smidgen of truth. The distinction between being better and being well is valuable. There's no reason why the notion of wellness should summon a vision of a Brave New World. It's based simply on the observation that, left to itself, clinical depression may go away but is then still liable to recur through life. Depression isn't a discrete, isolated state that, once banished, will always stay in abeyance. If there's a trigger for a recurrence, it can come back unless the mind is able to anticipate this assault and defend itself. With effective treatment, the agonies won't come again.

This is the sense in which a sufferer from depression realistically aspires to be well rather than better. The habits of depressive thinking can be remedied by psychological treatment, aided by medication, but it's better that they be cured. That's what happened to me, through sessions with a clinical psychologist, while my emotions were stabilised with drugs. I went to the psychologist initially in the desperate hope that she would say something – anything, though I didn't know what it could possibly be – that might ease the excruciating pain. In sessions stretching over several months, she stressed that our aim should be not just a return to normality but to attain a 'new normal'. The therapy had a specified goal, and we stopped when we felt we'd reached it. If I ever do slip back towards a depressive state later in life I will know what to do and how to counter it, and this to me is the equivalent of a cure.

Becoming well required abandoning certain trains of thought that were so ingrained and persistent that they were second nature to me. I'd laboured for many years under the immovable conviction that I was a failure in all the essential aspects of life that made it worthwhile, and that only serial disappointments awaited those who depended on me. Nothing but the trappings of professional success remained, and these superficialities mocked me by their relative insignificance.

This line of rueful thought didn't seem fatalistic but was what any rational person in my position would conclude. I'd long wondered why I found life difficult, believing that an answer could surely be uncovered if I thought hard enough while stripping away extraneous detail and scrutinising what remained. If only I could gain this knowledge, I would know how to fix the problem, whereupon a state of enduring contentment would await me and I would escape the inevitable scorn of those who looked to me. Yet there was no obvious pattern in these setbacks and founderings beyond the fact that they were mine and they were not what I'd imagined adult life would hold. The only conclusion I could draw was that things were destined to be that way and would never change. I would carry on, aware of the hollowness of my existence and a stranger to fulfilment and happiness.

The required shifts in my mental universe from this credo weren't radical, and they wouldn't have been effective if they had been, for they would have lacked credibility against my habits of reasoning. But the ability to make them while in a state of severe depression was a revelation to me. The psychological therapies that I experienced may have saved my life; they certainly changed it for good.

Suffering without treatment

In this book I give a non-technical but reliable account of current medical and scientific thinking on depression, and of what it takes to escape the hold of this affliction and become well. I aim to inform general readers about an under-reported and widely misunderstood subject. I also hope that sceptics, who believe that the right response to depression is for society to collectively develop a backbone, and sufferers from depressive disorders who despair of ever emerging from them will consider the evidence I set out and, respectively, think again.

I'm aware that, while depression is a widespread illness, my experience of it was atypical, to my benefit, because I got help and expert advice. Many other people suffer from depression without treatment or understanding, and they pay a heavy price in their personal relationships and in their work. I had friends and family who strove to support me through it and ensure that I came to no lasting harm. I had an employer prepared to back me without reservation even while I was, though formally present, incapable of fulfilling my contractual obligations. There were times when, as a full-time, salaried employee of a national newspaper, I would struggle to produce more than half a dozen words in a day, and even those wouldn't necessarily conform to the rules of English syntax and morphology. But I kept coming to the office, and colleagues kept discreetly looking away from the evidence of my incapacity.

I had the benefit of medical advice laid on by the company, and psychological therapy paid for through my employee health-insurance scheme. I had the right contacts through my work who could introduce me to leading figures in brain science, clinical psychology and other fields relevant to mental health. These scholars, psychologists, scientists, doctors and campaigners were prepared to talk to me freely and without any thought of reward, so that I could understand the fruits of their learning and research.

It inevitably sounds portentous for an author to say they were driven to write a particular book. Few writers are well known to the public. Few titles survive beyond an initial limited print run. But as my experience of depression was more fortunate than those of most sufferers, I genuinely felt a moral obligation to write about it so that others could learn of the same things that I found out. I knew little about mental illness when the events I describe started. I know enough now to be certain that those in the midst of major depressive disorder can be helped, that public policy needs to focus more intensively on mental health, and that providing evidence-based, cost-effective treatments is a priority.

This book is the result. It sets out what depression is and how it can be not only eased but overcome, personally and socially. The treatments evolve, but one day, perhaps not far off, a condition that in the prehistory of our species may have served an adaptive function but now brings only anguish and desolation can be lifted from humanity.

Oliver Kamm
London, July 2020

PART ONE

UNDERSTANDING DEPRESSION

1

WHAT IS DEPRESSION?

> I am now the most miserable man living. If what I feel were equally distributed to the whole human family, there would not be one cheerful face on the earth. Whether I shall ever be better I cannot tell; I awfully forbode I shall not. To remain as I am is impossible; I must die or be better, it appears to me.[1]
>
> Abraham Lincoln, 1841

There's no bleaker place than the human mind when it's unmoored from reason. I lost hold of mine at the tail end of 2013. The condition was diagnosed as severe clinical depression. It dominated my life, everything I did and thought in every waking moment, for a year while I sought, and fitfully managed, to piece my rationality back together.

The misery was remorseless, prolonged and agonising. It wouldn't abate for a moment. My unyielding, impassioned desire was for the pain to end. If I could have summoned the will, I'd have been prepared to adopt any means to secure that goal. For great stretches of that time, exactly covering the calendar from the end of one December to the end of the next, my overpowering thought, crowding out all else, was a fervent wish not to see another day. It was, I supposed, a type of insanity in which the mind no longer worked and the personality felt broken, in function and emotion. It was my first conscious encounter with depression, and the condition wasn't at all how I'd imagined it might be. Eventually, through the herculean efforts of others to reach me, I came out the other side.

My first hint that there was a problem came one day when I forgot my home address. I'd gone to pay for something in a shop, only to be told by the surprised assistant that I'd done it already. This was news to me, and it was a strange realisation. I had no recollection of an event that had happened only a few seconds earlier. I took it as a sign of unusual tiredness and that my mind was wandering, and decided to get a taxi home. That plan immediately fell apart. Normally I can recall obscure details of things I've read or seen, or names and faces from years past. I used to have a party trick whereby, if introduced to an apparent stranger at a social event, I'd be able to tell them the date and location when, years or decades previously, we'd met before. The information was always accurate and, though they wouldn't remember the occasion, it would tally exactly with their biography. Now I realised there was no purpose in hailing a taxi as I was unable to tell the driver where to take me.

I had no app then to tell me where I lived. Nor did anything in my wallet disclose the information, which proved surprisingly hard to divine by any other means. Years previously I'd been a contestant's 'friend' in the 'phone a friend' part of *Who Wants to Be a Millionaire?*, a once-popular television quiz show. I couldn't recall its origin, but that imperative dimly occurred to me now. Perhaps I should phone a friend to find out where I lived.

The quixotic nature of the scheme deterred me, and I took instead what seemed the sensible and only alternative course. I sat by the road waiting for the information to return, as it surely would in time. It didn't. In a line that likewise came to me unbidden but that I couldn't immediately place, it was curiouser and curiouser. I was fully awake but nothing would come to mind beyond immediate sensations: the traffic in front of me and spots of rain, amid an overwhelming sense of oppression, darkness and exhaustion.

That episode was the beginning of a long and precipitous descent into mental chaos. Or rather, it probably wasn't the beginning of the journey; it was the moment at which (so I later imagined) a primordial instinct forced a short-circuit of the mechanisms of thought, cognition and memory. The disorder had gestated long and slowly; like a pressure cooker that is unexpectedly opened, it now erupted and in unpredictable directions. I didn't understand what was happening or why, but a state of helplessness and fear became the governing principle of my existence.

Memoirs of mental disorder typically stress its mystifying character. My account is no different. Even to describe depression, let alone convey it, is difficult – not because for me the recollection is any longer afflictive, but because it's unlike anything else. Depression was by far the worst experience of my life. It grieves me that the later recollection of losing the ability to converse meaningfully with my mother, whose intellect towards the end of her life was lost to Alzheimer's, does not prompt the same anguish. I tried at the time, with deliberation, to regard her deterioration as the coda to our relationship, and was thus prepared for her death. There's no experience I've had, nor any I've seen, that plausibly compares to depression in the symptoms or the suffering. I've never felt such pain.

Physical hurt can be so intense as to be unendurable, but palliatives can dull it. Anaesthesia can remove the sensations altogether, though temporarily. With depression, there are no moments of relief. There's nothing, no induced state and no medication, that can temper it, let alone lift it. You wish only for oblivion. It's pervasive, seeping into your being, and it's hermetically encompassing, so you can't recall how things used to be or imagine how they could ever be otherwise.

For sufferers of severe depression, it's not just part of the scenery; it's the landscape. The agonies it brings are inescapable. They aren't blotted out even in fleeting moments of sleep, as they reoccur in a dreamlike state, in imagery elicited by the brain's activities, and then they flood back. When you wake up, there's a respite of maybe a few seconds before you recall that the world is not as you once knew it, in the time before depression, and then you recoil from the knowledge that you're still in it. The dismal realisation is compounded by the knowledge, the certainty, that the day ahead will be just like the one before, and the one before that. As Abraham

Lincoln suggested to his law partner in the epigraph above, death appears the merciful course in the absence of recovery.[2]

The reach of depression

Depression is a fact, not a phantom. Great minds in the sciences and humanities have studied the condition over millennia and investigated its causes, effects and remedies. Our understanding of depression is admittedly rudimentary, even after all this time, and theories that once seemed well established have turned out to be speculative or false. But partial knowledge is still knowledge, and sound science has guided the search for effective treatments. We don't entirely know how these psychological, medical and pharmacological interventions work but they do so predictably and not for arbitrary reasons. The range of remedies includes others that are outdated and dangerous, and some that are popular yet at best futile and potentially psychologically damaging. I had the latter type of treatment too, drawn from disciplines that verge on the pseudoscientific, before finding what worked – not just for me but, on the evidence of those clinical trials, for many others too.

Even allowing for the difficulty of distinguishing between intense conventional sadness and clinical illness, we know that depression is widespread. The World Health Organization (WHO) estimates that more than 264 million people of all ages currently live with depression.[3] Moreover, there is a demographic effect. According to the WHO: 'The number of persons with common mental disorders globally is going up, particularly in lower-income countries, because the population is growing and more people are living to the age when depression and anxiety most commonly occurs.'[4]

There is admittedly some dispute about whether the number of sufferers globally is stable or rising, but there is no doubt that depression is an immense social problem and pathology. An academic study in 2013 concluded that depression is the predominant mental-health problem worldwide, followed by anxiety, schizophrenia and bipolar disorder.[5] (Though depression and anxiety may overlap and exhibit similar symptoms, they are clinically distinct conditions. Depression is essentially a single illness whereas anxiety is an umbrella term covering a range of conditions, such as generalised anxiety or phobias and panic disorders.) Another study the same year cited depression as the second most prominent global cause of years lived with a disability, behind only lower back pain. In twenty-six countries, depression was the main cause of disability.[6]

In Britain, the Office for National Statistics has in recent years constructed data sets that go beyond standard economic measures of welfare, to include measures of wellbeing. It's a good idea, and consistent with what economists know about the limits of standard measures even as a narrow gauge of welfare. GDP, for example, is a measure of the total output of the economy, but it measures only market output. This leaves out, among other things, the immense area of unpaid labour in the home.

Wider measures of economic welfare are more subjective than growth in output or household incomes, which is one reason they are unlikely ever to supersede these rough-and-ready measures, but they do point to factors not captured by more conventional data. They indicate that in 2014 almost a fifth of the population of the UK aged sixteen or over exhibited symptoms of anxiety or depression, and the number was rising among women aged sixteen to twenty-four. There was also a significant split between the sexes, as the percentage was higher among women (22.5 per cent) than men (16.8 per cent).[7] Perhaps the difference was due to both the subjectivity of the measure and its reliance on self-reporting, and that men may be more reluctant to admit to problems. There's no immediate way of telling. And sceptics would point to the (apparently) suspiciously widespread incidence of mental disorder. Can it really be true that anxiety or depression affects roughly one member of every household in the country?

Well, it's not true at any one time (what's known as point prevalence), but it is true of lifetime prevalence, and, while striking, the finding isn't necessarily overstated. It doesn't tell us that Britain is now a nation with a 'therapy culture' which confuses sadness with illness. It refers only to *some* symptoms of depression or anxiety, rather than to the minimum number required for a medical professional to give a diagnosis of depressive disorder. And depression, contrary to the critics of the term, really is not a modern malady. It is a timeless and enduring affliction of the human mind. The terminology and diagnoses may change over centuries but the fact of it reoccurs.

Features of depression

If depression is a real condition, and it is, then what exactly is it? Far from being a grab-bag of grouses, depression has a precise clinical definition. The *Diagnostic and Statistical Manual of Mental Disorders* (*DSM*) used by health-care professionals in the US identifies depression as an episode of at least two weeks of a depressed mood or loss of pleasure or interest in almost all activities, along with at least five other symptoms, which cumulatively cause the sufferer significant distress and damage their ability to function socially or at work. The *DSM* has expanded its categories of mental disorder in the past generation, but that doesn't mean its designation of depression is arbitrary. Not at all: its criteria for diagnosing a depressive illness are carefully defined and extensive.

The *DSM*, published by the American Psychiatric Association, is the primary classification system used internationally in research. In the UK, researchers use the World Health Organization's *International Classification of Diseases* for clinical reasons but are less constrained by classificatory systems. Guidelines on clinical practice are published by the National Institute for Health and Care Excellence (NICE). Its conclusions on assessing and treating depression identify as principal symptoms a persistent sadness or low mood, and a marked loss of interest in other activities. The additional symptoms include disturbed (or excessive) sleep, rapid changes in appetite and weight, fatigue or loss of energy almost every day, an inability to concentrate or make decisions, sluggish or agitated movements, feelings of excessive worthlessness or guilt, and recurrent thoughts of death or suicide.[8] Again, this is a careful, clinical definition, based on observation and evidence.

Taking these criteria as symptomatic of depression, the NICE guidelines further distinguish between different states of the condition according to severity. These distinctions are what you'd intuitively expect. Mild depression exhibits few if any symptoms beyond the five required to make the diagnosis, and the sufferer has only mild functional impairment as a result. In cases of moderate depression, the symptoms and functional impairment are greater. And with severe depression, most of the symptoms are present and they substantially affect the sufferer's abilities to work and socialise. Sometimes these are combined with psychotic symptoms, but not always.

A psychotic disorder is a severe mental condition that causes sufferers to lose touch with reality (or, to put it more neutrally, they perceive the world around them differently from everyone else). Common psychotic symptoms are hallucinations and delusions. A delusion is a strongly held belief in spite of evidence that the belief is false (though I'm wary of using the language of mental illness in the realm of politics, some beliefs in conspiracy theories are hard to distinguish from psychotic disorders). But psychotic symptoms don't need to be present for a disorder to be classified as severe depression. Complicating the picture, a depressive episode can occur in the context of a recurrent depressive illness or as part of bipolar disorder. Based on their symptoms, these illnesses can look similar, but they will differ over time and require different treatments.

I had all the symptoms at once, including some that verged on the psychotic. They were all there, they were all severe, they all affected my ability to function in social situations or at work, and they caused unyielding distress. There were occasions when I heard voices and was unable to distinguish whether these were real or imagined. I marked them down as imaginary: I intellectually knew they were inside my head and not outside, and that nobody else could hear them. I couldn't have been mad, or the treatment I received would have struggled to make headway, but I feared insanity and the vividness of the experience unsettled me.

Mine was a severe disorder, but milder instances of depression are debilitating too. The technical name for mild depression is dysthymia. Sufferers liken it to a low-level, ever-present hum. You can ignore it and give every appearance of functioning normally, but you know it's there. It's far worse than an irritant because it's not temporary. To call it mild is in fact misleading, if not a misnomer, because it causes deep unhappiness and changes your behaviour. You become irritable, exhausted and more difficult for others to reach.

A thesis of depression: organic and treatable

Every case of depression is different and there are degrees of severity to the illness. There's no shortage of evidence, however, that the symptoms of depression are common to sufferers across societies and epochs. To the damaged mind as it looks outwards, the world in its entirety is transformed into a slough of despond. To a person racked by severe depression, every street, every building and every room is a further entrance to despair. It's a natural urge for a sufferer to try any expedient to stop the pain, and it's tempting to conclude that, as so much is unknown about the mind, whatever works for the patient is valid and should be accepted. But that can't be right. The details matter. We may not know how various treatments for depression affect the brain, but researchers are able to employ statistical techniques to measure their impact. If the sample size is too small and the study lacks controls and follow-up, then – whatever it feels like to the patient – the treatment lacks scientific support. There are a few treatments for depression that pass this test and very many that don't.

Depression may be common but it is still not widely understood; hence the widespread confusion about attempted remedies. Treatment for mental disorder even in an affluent society like Britain is patchy in quality and geographical availability. The reliability of publicly available information is, sadly, not enhanced by the attitudes of those, including some ostensibly sophisticated commentators in my own profession of journalism, who dispute that depression exists or who niggle about the term's applicability. Conversely, the prospect of recovery from depression's many manifestations, and especially its most severe forms, is scarcely imaginable to those who are immersed within them.

Some inquiries by medical practitioners and psychologists have taken wrong turnings. The field of mental illness also has an ample share of dogmatists who spin theories that lack substantiation, and of quacks who devise and practise remedies that are at best no better than allowing the passage of time to heal. Mostly these folk interventions are useless but sometimes they're damaging, delivered with enough of a patina of training and knowledge to be plausible to people in desperate need. And if at the bare minimum these ineffective techniques persuade sufferers that they're already getting help and should thus forgo the better ones, then that is in itself an act of harm.

While our understanding of depression is as yet limited, I'm confident of this thesis:

- Clinical depression is an illness like any other, with a complex mix of causes that are physical, psychological and social.
- It is a disorder of a physical organ, and the most complex one we have: the brain.
- The nature and causes of depression will one day yield to scientific inquiry.
- There are already treatments available that work – though not always, and not for everyone, and it's not clear what the mechanism is by which they provide relief and ultimately recovery to the distempered mind.
- These methods and medicines will be refined, and still better approaches discovered, as scientific understanding of the brain advances.

You needn't take my word for any of this. I'm not a neuroscientist, psychologist or psychiatrist. I have no training in these fields. I'm a journalist who became, in both senses of the term, an interested party. But I can point to the accounts of many others, and the depiction of depression in works of the imagination, demonstrating that the condition is a constant of the human experience and has devastating effects on people of all ages and in all walks of life. And I can recount the state of scientific knowledge of depression and the routes by which it's come.

Full recovery from depression may happen on its own but this isn't true of all sufferers, and, even for those whose condition is tractable, the prospect will be tragically too late for some who believe they are doomed to always carry the illness with them. The WHO estimates that each year almost 800,000 people globally take their own lives and that depression is the major contributor to these deaths.[9]

In this book I want to explain what depression is like and how it's possible to recover from it. My account on its own is just one story, but the treatments I was fortunate enough to get have been demonstrated to work, through clinical trials and statistical analysis, not with everyone but certainly with many. If you are in utter despair or find it hard to cope with the travails of someone close to you who is in that state, there is help and the promise of emerging from the darkness. There is always help.

2

WHAT CAUSES DEPRESSION?

> It ought to be generally known that the source of our pleasure, merriment, laughter and amusement, as of our grief, pain, anxiety and tears, is none other than the brain. It is specially the organ which enables us to think, see and hear, and to distinguish the ugly and the beautiful, the bad and the good, pleasant and unpleasant.[1]
>
> Hippocrates, c.400 BCE

Depression is not, as some critics suppose, a recently confected and fashionable label to describe life's standard regrets and woes. The depressive state is coterminous with humanity. It sometimes erupts at a particular stage in life. And this is how it happened to me. The trigger points were not obvious at the time, but they were there.

My depression had all the main elements listed in the diagnostic manuals. An alteration of mood and thought affected my appearance and behaviour. I became thin, drawn and haggard. I moved slowly and haltingly. I showed scant awareness of my surroundings or of anyone who sat with me. I'd speak, if at all, only when directly asked a question, and then briefly and without apparent awareness of what had been said before. Tearfulness could come at any moment, wherever I happened to be. The changes were marked and immediately noticeable to friends and colleagues. And only in retrospect did it occur to me that the descent had been prolonged and the behavioural quirks observed by those I loved. The depression simmered over years before erupting into a continuous anguish.

There was no definable trigger for this state other than the vicissitudes of life. My difficulties were common and have been known by many others too, but they were experienced by me with peculiar vividness under a carapace of stoicism while I confided in nobody. Even my closest friends had barely seen me for around a decade. In that time, before the depression struck, my professional persona was fulfilling and ostensibly successful, but my personal life felt like a long catalogue of stumbles. The gulf weighed on me. A caring but unsuitable marriage and separation produced loneliness for both of us. Bereavement, the death in 2011 of a father I loved and realised I'd insufficiently spoken to, compounded a sense of failure and of forever falling short. I brought up two young children on my own and found it hard.

It felt as if life had passed by me when I wasn't looking. I was buffeted by eddies of daily existence that I couldn't control, and I came to believe that those who looked to me for care and comfort were sure to realise my inadequacies.

The idea of a midlife crisis is often dismissed as hackneyed and self-indulgent, and it had always seemed so to me. I'd not expected to go through one and I didn't recognise it as such when it happened. I'm still not sure this was what it amounted to. The fatalism of such a notion didn't repel me so much as make it empty. We all have disappointments and setbacks, and – I told myself – a purposeful life involves navigating them and sometimes stoically accepting them. If I'd thought of the travails of middle age at all, it was as an unconvincing paradigm of life, for the opportunities are always available to counterbalance them.

My readings as a political commentator told me that I had to accept this inevitability. Michael Oakeshott, among the great political theorists of the twentieth century, stressed that in public life we need to acclimatise ourselves to the world as it is rather than as we might wish it to be. He wrote of the change of perception as the solipsism of youth gives way to the responsibilities of adulthood: 'For most there is what Conrad called the "shadow line" which, when we pass it, discloses a solid world of things, each with its fixed shape, each with its own point of balance, each with its price; a world of fact, not poetic image, in which what we have spent on one thing we cannot spend on another; a world inhabited by others besides ourselves who cannot be reduced to mere reflections of our own emotions.'[2]

My philosophical inclination is more optimistic, but this is an essential truth. And I have no fear of the passage of time. It was Oakeshott's scepticism, and his suspicion of absolutist claims to have found answers to enduring human conundrums, that appealed to me as an approach to life's quandaries. Yet when I crossed the shadow line, I found a world that was not as Oakeshott had written. It was unfamiliar and lacked the clear lines of navigation that I'd imagined. It wasn't a void, as the paths I'd embarked on were still there and so were the people I was close to. But the way ahead was indistinct and tortuous. Instead of entering a world of fact, and gaining the confidence born of knowing how life worked and what its purposes are, I was shrouded by doubt.

Like many single parents, I had no idea how to raise young children. Beyond wishing them to flourish, or at least for them to come to no harm, I was clueless. To partially compensate for the fear that I was failing them, and assuming that it would benefit them in the longer term, I took the bizarrely counterproductive step for a couple of years of working furiously hard. It was callous, and it appeared so to me in retrospect, but I was confident I knew how to do this side of life through everything, even when I was distracted and the work wasn't of the best quality I could manage. I had the ill-formed belief that if I could only be confident that no obstacles would ever arise in performing my professional and public role, it would enable me to concentrate on raising the standard of my childcare. It was a ridiculous notion. It was in fact displacement activity and it should have been obvious to me that I would later worry about neglecting my parental duties; and so it turned out.

Depression and the voyage of life

Amid these apprehensions and confusions, while apparently secure in my position in life, I sat one day chatting to an old friend, the writer Christopher Hitchens, in his flat in Washington DC. Christopher was unusual as a British journalist (though he had lately acquired US citizenship in a stand of solidarity after the 9/11 terrorist attacks) in having become an established and admired commentator within the Beltway. He'd had cancer of the oesophagus, in an advanced stage, diagnosed several months earlier. As the afternoon wore on and twilight fell, and he didn't turn on the lights, he said of the experience of having friends call on him: 'There's always the lingering thought, left in the air, of whether this is goodbye.'

For us, I knew it was goodbye. There was no question about it. He was too frail ever to travel to Britain again and he'd gone to immense pains, physical and metaphorical, to see me to be interviewed for my newspaper. That was the purpose of my visit. The word 'tragedy' is overused and often misapplied to either mere misfortune or malign human agency, but here it made complete sense. The tragedy was before me in person. Christopher's eloquence as a commentator and debater was unmatched and he was departing life with much, very much, still to say. As an atheist, he was assailed but characteristically unperturbed by messages from devout (and usually anonymous) correspondents declaring his plight to be divine judgement. He saw no cruelty in his situation, but rather the workings of an impersonal universe.

He wrote a few weeks after the diagnosis: 'To the dumb question "Why me?" the cosmos barely bothers to return the reply: Why not?'[3]

There was likewise no existential wondering for me in knowing we wouldn't meet again. He would soon die, in conditions that I hoped would be painless and peaceable, and become dust. (As it turned out, he left his mortal remains to science. There was no funeral or cremation.) But he was sixty-one and would not see his children married. Never to experience a rite in family life, rather than vacating the world of ideas and polemical exchanges he was famed for, was his acutest regret. Mine, more prosaically, was to be losing a fixture in my life: a friend whose wisdom and conviviality I'd relied on would shortly be extinguished for eternity and, more pressingly, for all the time remaining to me.

In the National Gallery in Washington hangs a massive four-part allegorical work called *The Voyage of Life* by Thomas Cole, painted in 1842. I happened upon it for the first time that weekend when I was awaiting word from Christopher on whether he was feeling strong enough to chat, and spent a day walking along the Mall. It's a series of landscape paintings that on first impression convey the sort of religious piety, with a traveller guided through life to an ultimate heavenly retreat, that repels the rational mind. The least convincing reason for embracing theism is the promise of eternal life, as religious thinkers through the ages have been uniformly unable to depict why it should be desirable. Lloyd George said, with far greater plausibility, that as a child he'd feared heaven more than hell because the prospect of being trapped in an eternal church service without possibility of escape was too awful to bear.[4]

Yet a closer look at the paintings shows a different focus. In each part of the quadriptych is a traveller, the same traveller, in a boat but at different ages. The first, *Childhood*, shows a baby exultant at having emerged from the darkness of pre-existence. Behind, with a hand on the tiller and an outstretched arm, is the child's radiant guardian angel, guiding the boat across the placid waters of a stream surrounded by lush vegetation. At the prow of the boat is the figurehead of an angel bearing an hourglass. All is promise. The second, *Youth*, depicts the child as a young man, full of hope. He waves in gratitude and anticipation to the guardian angel, who is on the bank rather than in the boat, as he embarks on adulthood. Towering in the distance, above the clouds, is a celestial palace. But what is invisible to the traveller, while visible to the viewer of the painting, is the winding and tortuous route to that heavenly retreat. The third, *Manhood*, is the darkest of the paintings for it shows the voyager in maturity, bearded and sturdy – yet helpless as the boat, battered and swept down by the currents, seems to head to imminent disaster. The rocks crowd in and the voyager looks heavenwards, with hands clasped in prayer and supplication. He can't see it but, very distant and in the background, through a gap in the storm, is the angel still keeping watch. And the fourth, *Old Age*, shows the reward for constancy. The voyager, now white-haired, has arrived at calm waters, as the heavens part and light shines down, while the guardian angel directly ahead points to the glories that await.

The midlife dip

The whole scheme of this massive work of landscape painting is remote from modernity. When I saw it, I thought of the remorselessly saccharine artworks of evangelical tracts and instinctively recoiled. How wrong I was. The work is an expression in the language (literally the language, as the artist wrote an exposition of his highly programmatic painting) and imagery of its time of an enduring aspect of life. As we become independent, we make life choices. Some of these seem workaday but have momentous effects. Sometimes we postpone and fail to take a course of action for fear of its consequences, and sometimes we make decisions we later regret. The voyage of life, however, isn't just a lament for what might have been. It has observed regularities in its stages. Psychologists have considered why the experience of dissatisfaction and turmoil in middle age is so common, and economists have studied the conditions for happiness. In an influential paper in 2008, the economists Andrew Oswald and David Blanchflower presented 'evidence that psychological well-being is U-shaped through life'.[5]

Why this should be so is not dealt with at any length in the social-science literature. There is in economics a finding known as the Easterlin paradox (named after the American economist Richard Easterlin) that once a developed country crosses a certain threshold of average income, growth in GDP per head doesn't translate into greater happiness. But the dip in the life cycle, for which the empirical findings are strong, isn't satisfactorily explained. Oswald and Blanchflower point to one theory, proposed by the psychologist Laura Carstensen, that (as the economists summarise it) 'age is associated with increasing motivation to derive emotional meaning from life and decreasing motivation to expand one's horizons'.[6] It's plausible. What lies behind the U-shaped happiness curve is an issue that needs study, not least because of its relevance in the treatment of mental disorder across ages. Though I can proffer no explanation of the U shape beyond what's in the literature, it describes the experience of many people I've talked to. And it is mine too.[7]

For all the conventional waywardness of the middle period of life, I'd assumed that the anxieties of mine were cushioned. In my mid forties I started a new career, in journalism, that I was fortunate to be offered and was well suited for. And as time went on, I had the unlikely and unanticipated good fortune to meet the woman of my dreams. She was a single mother with two young children. We talked, on and on, and it was a delight to me. We fell intensely in love, were fascinated to follow the tributaries of each other's thoughts, and seemed to have found our personal refuges.

Yearnings of the heart undo our sense of order. They may in themselves be a form of irrationalism, if not of mild insanity. They are prompted by an initial and delirious rush of feeling against which resistance is powerless. But they are not fated to bring happiness; often enough, they cause misery. The lover whose passion is unrequited is the quintessential sufferer. W. B. Yeats asks in 'The Tower': 'Does the imagination dwell the most / Upon a woman won or woman lost?'[8] The answer is obvious, and the poet knew it well. Hence the addressee, 'impatient to be gone', doesn't even trouble to state it.[9]

Yet for love to bring happiness requires more than that it be accepted and reciprocated. The initial ardour is not enough. However prosaic the reality may sound, it needs to be channelled and tamed by the forces of reason, by the knowledge that what the heart desires accords with an appreciation of the real and not imaginary qualities of the loved one. And this is hard to understand at the time. I've seen marriages grow stronger because the knowledge of each other is fascinating to both. My father's second marriage was like this. The last time I saw him, which was the only occasion I can recall when he looked fully his age, he was stricken with cancer and close to death. My stepmother said as we left the hospital that her evident distress was the price you pay for love. It was an example of a closeness I knew from observation but had never experienced myself. Love transmutes infatuation through its own alchemy, so that desire is entrenched by esteem. It can only survive, or at least only healthily so, if it's well grounded in knowledge.

And in my love, I messed up spectacularly. She, the object of my thoughts and all my desires, persevered but was driven to distraction and despair by my seeming distance and tardiness to commit, and reasonably concluded that I was not the quality of man she'd assumed. After three years we parted, perplexed. For me, the knowledge of my failure compounded the sense of loss. There had been, in her estimation of me, no divergence between the emotions and the intellect: she had believed in me, and those beliefs had proved mistaken. Instead of being her sanctuary, I had proved to be her Casaubon: a man, as Dorothea comes to understand of her beloved in George Eliot's *Middlemarch*, of irredeemable mediocrity rather than creativeness and compassion. The realisation appalled me, yet the facts were stark and inescapable. Her very words had been 'I believe in us', yet she had been forced to conclude on the strict evidence presented, and on nothing else, that I was unworthy of her. It was true, and I shrank from the knowledge of it.

In that same summer, my children finished primary school and left to live overseas. There they didn't flourish, and I worried. I kept telling myself I'd done my best to prepare them, and recalled that we'd had happy times amid the disruption, but these rationalisations failed to convince me. Commuting long distances to work each day, I'd frequently had my travel frustrated by an unreliable rail service and I relied again and again on the indulgence of a child-carer staying late or of the local mothers' network. Often I'd sat on the veranda of the cottage where we'd moved to and watched the children in the garden. Every parent has the sense of a child's innocence and vulnerability, and I lacked the power to protect them; I had a recurring mental image of clouds gathering beyond as, unawares, they played. It was a very compressed vision of *The Voyage of Life*. Within me, self-reproach hatched, incubated and grew to execration.

Grief and sadness are part of living, and so are their eventual softening amid the solace of friends, work and recreation. Knowing this, I waited. Yet the darkness didn't dissipate. It thickened. With it, my behaviour became unpredictable and my work was of increasingly chequered and erratic quality. My journalism hinges on the stuff I know: the minutiae of recondite subjects, literary and historical references, and the footnotes of forgotten authors. Now I was having difficulty recalling things I'd been told only a few minutes previously. My writing had become excruciatingly slow and was strewn with grammatical errors. My friends were alarmed. I'd send them barely coherent emails in the middle of the night. Trying to make sense of the communications, they'd patiently respond when they awoke and then call at my home in the evening to discreetly check on me.

It was at that stage, and in that state, that memory departed altogether as I sat by the roadside. Failing to recall where I lived was a startling descent and it was my first inkling there was a problem. I didn't know what it was and in a sense I still don't. It had never happened before and I was perplexed at the detachment of my mind from what I willed it to do. How the mind works is one of science's remaining intractable questions. We know that genes control brain development, and that neurons firing within the brain produce consciousness, but we don't know the mechanisms that produce thought. Perhaps my mind suddenly shutting down was its way of coping with a condition that had built up over a long time, but I can only guess. All I can confidently say is that, while there were environmental triggers for depression, the cause of it was internal. Every cognitive state comes down to the workings of the most complex organ we possess, but beyond this we know little.

The mysteriousness of mental states

Modern neuroscience (the study of the brain), cognitive psychology (the study of mental processes) and philosophy of mind (reflection on mental states and especially the relation of the mind to the body) all converge on the conclusion that depression must be physiological in origin, but it's otherwise mysterious.

The problem in identifying depression and other mental disorders is that they are much harder to recognise than physical disease, as there is no brain-imaging machine or scan that can reliably diagnose them. The brain of someone with depression looks essentially the same as the brain of someone without it. The caveat here matters, though: they look *essentially* but not *absolutely* alike. There are some suggestive differences. As the clinical psychologist David Clark and the economist Richard Layard have noted: 'Most depressed people differ from other people in two ways. First, their amygdala is over-reactive (this is the structure deep in the brain which for example promotes the fight-or-flight response). Secondly, much of their pre-frontal cortex is underactive (this is the "more conscious" region which normally regulates our emotions and reactions).'[10]

Depression is the outcome of the same brain circuitry that everyone has, but the circuitry may work slightly differently in a depressive state. I stress in later chapters that we're not able to identify as the *cause* of depression particular differences (in, say, brain states or chemical balances) between depressed people and neurologically typical people. Those differences could just as easily be *symptoms* of depression. But even if we only know depression from its symptoms, we can still conclude with assurance that these are real.

The recognisable signs of mental illness are recorded in literature since ancient times. Hippocrates, the Greek physician popularly known as the father of medicine, knew well of these disorders. His name has been invoked through the ages, in the form of the Hippocratic Oath to abstain from doing harm, as being synonymous with the ethic of medical care. Hippocrates dismissed the notions that illness was caused by supernatural possession and that it could be cured by invocations to the gods. With remarkable percipience and in accord with what became the findings of modern science, he identified the brain as the seat of consciousness. Hence mental abnormality was, he surmised, due not to evil spirits but to disturbance within the brain. He wrote: 'It is the brain too which is the seat of madness and delirium, of the fears and frights which assail us, often by night, but sometimes even by day; it is there where lies the cause of insomnia and sleepwalking, of thoughts that will not come, forgotten duties and eccentricities. All such things result from an unhealthy condition of the brain.'[11]

The fears and frights are never the same. They're always specific to the person who labours under them. But the symptoms are distinctive and the diagnosis of them is common over centuries. We know this because of the way that artists, poets, dramatists, novelists and perfectly ordinary people have described them. These accounts are strikingly similar.

The condition has gone under different names, but it is included in every case in what we now call clinical depression. To realise that this state is intrinsically human, that it's been known in every generation and is illuminated in the recorded experiences of others, was for me a step towards understanding and dispelling those frights.

The missing theory of depression

If depression is a strange experience for a sufferer to undergo, it's also a notoriously hit-and-miss affair for the mental-health professional to diagnose and treat. Alighting on the right remedy is often a process of trial and error to find what works. We don't know enough about depression even to understand what's happening when the sufferer finds relief from the symptoms. That's why I refer to the condition as mysterious. But what we do know is reliable and is grounded in evidence. We actually have a lot of evidence, drawn from randomised controlled trials, of how to treat depression. Our understanding of mental illness is hampered more by a lack of theory than of data.

That may sound surprising. Typically, we think of a theory as something tentative and unproven, whereas science is about hard evidence. In fact, it's not quite like that. Take a fanciful example: my theory of why John blushes a deep crimson colour when Mary approaches is that he has an undeclared crush on her. This hypothesis may be true (it probably is) but it's still just my hunch. I don't have any evidence beyond this observation to support it. A 'theory' in the scientific sense isn't like this: it is a statement of general laws that explain the evidence of something. Hence we talk of the *theory of gravity*, which says that all objects with mass attract each other according to a strict mathematical relationship involving the distance between them. The *theory of relativity* advances propositions about the speed of light and the curvature of space-time. The *theory of evolution* says that change in the characteristics of species over many generations is the outcome of natural selection and random mutation.

In the past 150 years, the treatment of physical illness has been revolutionised by the *germ theory of disease* – along with discoveries such as X-rays, which have greatly assisted the practice of surgery. Depression has effective treatments based on clinical evidence. We know these can help people with moderate or severe depression and that they work better than placebos (that is, 'dummy medicine' with no active substance though sometimes with a therapeutic effect) for sufferers, as they did for me. The tests and the statistical analysis have been done. But there the parallel with physical illness runs out, as our reliable knowledge of mental disorder is more tenuous than our knowledge, drawn from germ theory, of how micro-organisms invade the body, infect wounds and cause disease.

The explanation for mental illness has proved intractable beyond the accuracy of Hippocrates' thesis. We don't know anything like enough to judge what causes depression and other mental disorders. It's clear, though, how these conditions will eventually be understood. Our knowledge will expand through scientific study of the mind and of the medical, pharmacological and psychological treatments that are still in their early stages but are demonstrably effective. Science and reason are our route to understanding the material world, and this includes the life of the mind. These approaches are not only indispensable: they're all we have to make sense of the external world. Let me explain the grounds of my confidence that they are the route to understanding and dispelling the devastating afflictions of mental disorder.

Humans wonder. It's what we do. Our mental capacity is without parallel in time or place. No other species through geological ages has managed to understand the world around us. We've gained that knowledge through critical inquiry, not from the fruits of revelation. Reflection doesn't lend itself to dogma; there's always the possibility that our conclusions about the nature of things will be overturned, or at least overtaken, by new discoveries. Liberal civilisation is averse to claims of certainty. Hence it's easy to discount just how vast is the field of reliable knowledge. The permanent possibility, however remote, of error makes scientists look askance at words like *certain* and *indisputable*, yet for all practical purposes these adjectives do apply to our knowledge of atoms or genes or the evolution of organisms. Nothing will diminish that body of knowledge. All that will be discovered about the external world, for the rest of time, will be consistent with the known laws of physics, chemistry and biology.

But there are two areas of scientific exploration where what we know is so partial as to be minuscule: what's *out there*, in the rest of the universe; and what's *in here*, the stuff of mind.

On the second of these questions, you'd imagine the scope for understanding is greater, because the physical dimensions are much, much smaller. We're talking, after all, not of the expanses of deep space but of an object that's about the size and texture of a cauliflower: the human brain.

Yet, while the brain is the most complex entity we know of, our understanding of it remains slight. It's a physical organ containing around 100 billion neurons (or nerve cells) along with a vast number of connections between them. Neuroscience is still in its infancy. Only in the last century did scientists discover that neurons are the building blocks of the brain and that the connections they make produce perception, cognition and emotion. And because our knowledge of how the mind works is limited, superstitions concerning mental health and other aspects of psychology and psychiatry abound. Some of these have a very long history. There remains a stubborn conviction even among highly educated people, including some who are trained in medicine and psychiatry, that, because we're not able yet to come up with a full explanation for mental illness, supernatural influences may in extreme cases be at work.

That's a salutary reminder that there is nothing historically inevitable about scientific discovery and the expansion of knowledge. In a fine short story titled 'Forms of Things Unknown', C. S. Lewis depicted astronauts on the moon encountering Medusa and being turned to stone by her gaze. The scientific understanding ought to have consigned myth and legend to literature. Yet almost 2,500 years after Hippocrates lived, his assessment that diseases of the mind are disorders of the brain rather than the work of evil spirits remains a truth that is contested. My optimism that depression will eventually be understood is predicated on the defence of science. If we abandon science and embrace notions that are unevidenced, we won't get anywhere. We'll be stuck with the understandings of a pre-scientific era, which are proving hardy even after centuries.

Depression and the demonic

Though refuted by Hippocrates, the notion that mental disorder was caused by evil spirits, or alternatively was a divine punishment, has long outlived him. It was for almost all of recorded history an apparently serviceable explanation for illnesses of mind and body, and was almost universally believed. It may be that remedial actions in prehistoric times reflected this belief too.

Skulls from the Neolithic era, from around 10,000 BC, have been excavated in Europe that show evidence of trepanning. This is when a surgeon drills a hole into the skull while the patient is still alive. We can't know the reason for prehistoric trepanning, but it wasn't limited to Europe and has been practised in more recent times. A skull fragment was found in 1865 in Cuzco, the centre of the Incan Empire, in Peru by the American archaeologist Ephraim George Squier. It dated from between 1400 and 1530 AD and had a large and regular rectangular opening. Squier's explanation of this cranial fragment was that a battle wound had broken the skull and that this had affected the warrior's behaviour. The surgery was undertaken not only because of behavioural symptoms but because the skull had already been punctured. We can't know, but the pain and disturbance must have been extreme to elicit such a radical procedure. It's possible that the anguish of depression was among the maladies attributed to the spirit world.[12]

What is purely speculative in archaeological finds becomes suggestive in sacred literature. Scripture abounds with accounts of demonic possession, and it's important not to obscure their theological message by attempting to explain them away in naturalistic terms.[13] Even so, the account in Mark's Gospel of the exorcism of the Gerasene demoniac describes behaviour that to a modern reader is redolent of mental disorder yet is attributed to occult forces. He was 'a man with an unclean spirit, Who had his dwelling among the tombs; and no man could bind him, no, not with chains: Because that he had been often bound with fetters and chains, and the chains had been plucked asunder by him, and the fetters broken in pieces: neither could any man tame him. And always, night and day, he was in the mountains, and in the tombs, crying, and cutting himself with stones.'[14]

We can't strip scripture of its legends and then imagine we're penetrating to a historical narrative beneath it. Biblical criticism doesn't work like that. Instead, we have to try to infer the sense that a biblical account made to its authors and audience. It may be that the demoniac never existed. Yet the symptoms that the gospel account describes are characteristic of mood disorder, including the practice of self-mutilation and the wish to withdraw from society. To attribute this to demonic possession was a common diagnosis of the time, as two medical historians have noted: 'Mental illness was generally considered among the pathologies caused by demonic possession, which included epilepsy, common sins like lying or stealing, and even the ability of predicting the future. Demonic possession was therefore both an explanation and a solution for an unacceptable behavior in society.'[15]

Medical writers through the Middle Ages and as late as the seventeenth century were concerned to explain disease with reference to the workings of the occult. The mistaken beliefs of other eras invite the condescension of posterity, and some historians have amply provided it. In his study of the European witch mania, Hugh Trevor-Roper managed to insinuate a touch of misogyny as well by referring to the 'mental rubbish of peasant credulity and feminine hysteria'.[16]

Yet the mental universe of these times was different from modernity and we must judge it in its own terms. In his seminal study of this period, *Religion and the Decline of Magic*, Keith Thomas showed that belief in magic was not an idiosyncratic or primitive phenomenon in Early Modern England (lasting from around 1500 to 1700). Nor was it the exclusive preserve of the peasantry. Rather, it was part of the intellectual mainstream, in which royalty too was fascinated by witchcraft and guided by astrology. Thomas notes that King James I 'regarded academic medicine as mere conjecture and therefore useless', and that many of his learned contemporaries took the same view. It is hence 'not surprising that supernatural explanations of mental depression were advanced or that the main psychotherapists were the clergy. Physic alone was not enough to cure melancholy, declared the Puritan oracle, William Perkins.'[17]

There is a remarkable resource showing this facet of seventeenth-century intellectual history in the writings of Richard Napier, an Anglican priest and physician who was a contemporary of Shakespeare. Napier practised medicine in Great Linford in Buckinghamshire from 1597 to his death in 1634, and some sixty of his casebooks survive. He was an Oxford graduate, attuned to the discoveries of the age, yet he made no effort to dissuade his parishioners from believing that their physical and mental ailments were due to witchcraft or demonic possession. If he demurred at all it was to refrain from assenting to specific claims of sorcery by named people rather than to cast doubt on the reality of witches, demons and fairies. He was fascinated by alchemy, the quest to turn base metals into gold. His education, vocation and Anglican credo were compatible with all of this. The occult, drawn in Napier's case from immersion in neo-Platonic and Hermetic sources, was part of the belief system of the educated as well as that of the wider populace.

Those who sought Napier's help were suffering from mental disorders as well as physical ailments. It fell to him, as a clergyman ministering to his flock, not only to listen but to heal. A meticulous study[18] of the social history of insanity in this period reveals that Napier treated more than 2,000 psychologically disturbed patients, including 264 who believed themselves bewitched. Most of these patients were women. This sex imbalance was perhaps due to a folk belief that women were more likely to be invaded by malevolent spirits, though there was also a widespread popular belief in succubi, namely female demons who would have sex with men while they slept. In any event, Napier's remedy for mental disorder was exorcism. For those who were 'mopish and distempered in brain', he exhorted the supernatural forces that assailed them: 'First let them blood . . . then say, "Lord, I beseech Thee, let the corruption of Satan come out of this man or woman or child that doth so trouble or vex her or him."'[19]

As one recent commentator, Andrew Scull, has observed on this treatment of mental disorder: 'It was an eclectic mix of magic, religion, supernaturalism and medicine that seems to have matched the beliefs of both the learned and *hoi polloi*, who thought that these realms could and indeed must be reconciled by those who sought to influence the course of a variety of disorders.'[20]

The coexistence of science and magic

Up to the late seventeenth and early eighteenth centuries, and even among the greatest of intellects, science coexisted with superstition. The starkest single case is Isaac Newton. As a young man, Newton invented calculus, discovered that the same force of gravity that causes apples to fall from a tree to earth also guides the path of celestial bodies, advanced the new science of reflector telescopes, and demonstrated that white light was a mix of coloured rays that could be distinguished when refracted in a prism.

All of these individually, let alone cumulatively, rank among the greatest intellectual achievements of western civilisation. Yet the writings in which they were expressed testify to another side of Newton's intellectual universe. His *Opticks* (1704) applied his discoveries to explain how raindrops refract sunlight to produce a glorious spectrum. Rainbows had long been understood as a sign of God's covenant and assurance that after Noah there would be no universal flood;[21] they were in fact explicable by purely natural means. Look closely at Newton's reasoning, however, and you find that he is determined to demonstrate that his findings point to the workings of the divine: 'For so far as we can know by natural Philosophy what is the first Cause, what Power he has over us, and what Benefits we receive from him, so far our Duty towards him, as well as that towards one another, will appear to us by the Light of Nature.'[22]

It was a form of the teleological argument for the existence of God: the notion that the evidence of design in the universe testifies to a creator. The best-known version of this argument was advanced a century later by the Anglican divine William Paley in his book *Natural Theology* (1802), drawing an analogy with a watch. The intricacy of the watch's design points to a watchmaker, and likewise the design of the natural world demonstrates the existence of God, who made it. In between Newton's and Paley's expression of this idea, the philosopher David Hume destroyed it; Paley was apparently unaware of this (as are numerous authors of Christian apologetics to this day).[23]

But Newton went much further than arguing that the complexity of the natural world is evidence of purposeful design by God. He was determined to demonstrate the workings of the divine through complex numerology (or number symbolism) and thereby calculate the exact date of Christ's return to earth. On his death, he left a manuscript on biblical prophecy running to 850 pages. And that's not all: like Napier, he was devoted to the practice of alchemy. His unpublished writings on this subject run to more than a million words. His writings on prophecy were still more extensive, and both subjects occupied him to a far greater extent than physics and astronomy.

The oddity of this juxtaposition has been commented on by science pundits and practitioners. According to Martin Gardner: 'There are two reasons for viewing Newton's career as awesome: his stupendous discoveries in mathematics and physics, and the equally stupendous stupidity of his theology.'[24] I'd prefer to say that the theology was heterodox and eccentric even by the standards of the day. Yet it is a great historical conundrum what more Newton might have achieved had he devoted himself to his scientific endeavours. Extraordinarily, he regarded his theological and alchemical quests as part of the same discipline of critical inquiry. And, while an extreme case of brilliance allied to superstition, and of the stupendous industry involved in explicating ideas, he was still typical of his time in holding differing world views simultaneously.

*

If Isaac Newton, the most learned figure of his age and perhaps of any age, believed in magic, it is scarcely surprising that folk beliefs were strong among the common people. Belief in the supernatural persists in all cultures and across different traditions; perhaps more surprisingly, so does belief in the demonic origin of disturbance. An article by an academic psychologist notes: 'In Western societies demonology survived as an explanation of mental health problems right up until the eighteenth century, when witchcraft and demonic possession were common explanations for psychopathology. Nevertheless . . . demonic or spirit possession is still a common explanation for mental health problems in some less developed areas of the world – especially where witchcraft and voodoo are still important features of the local culture such as Haiti and some areas of Western Africa.'[25]

This is surely right. But the same is true of secular and pluralist western societies too. A British GP called Thomas O'Brien was struck off the medical register in 2015 after the Medical Practitioners' Tribunal Service found him guilty of serious misconduct, concluding that he was a 'risk to the public' and that he had 'repeatedly breached professional boundaries'. O'Brien, a devout Christian, had taken a patient who suffered severe clinical depression to his church to perform an exorcism and told her she would be cursed if she reported him to the medical authorities. He urged her to discontinue taking antidepressants and told her that psychiatrists do 'the devil's work'.[26]

It's an extreme case of the abuse of authority, but the underlying conviction that mental disorder has an occult origin is not rare in western societies in the twenty-first century, and is held even among medical professionals. I was astonished, after my experience of depression, when one of the leading newspapers in the English-speaking world, the *Washington Post*, carried an opinion piece by Richard Gallagher, a Professor of Clinical Psychiatry at New York Medical College, explaining how he helps to spot the symptoms of demonic possession. It's not metaphorical. That's what he really imagines he does. The article gives an example of a subject who spoke in tongues, including Latin, that were completely unfamiliar to her outside her trances. Did Gallagher himself hear this? Predictably he did not; he relied on the testimony of others, while concluding: 'The same habits that shape what I do as a professor and psychiatrist – open-mindedness, respect for evidence and compassion for suffering people – led me to aid in the work of discerning attacks by what I believe are evil spirits and, just as critically, differentiating these extremely rare events from medical conditions.'[27]

It's salutary to note how thin and porous the boundary remains between folk belief and the rationalisations of ostensibly sophisticated people, even in this case a specialist in mental health. Yet we *must* discard supernatural explanations for mental disorder and for anything else that happens in the world. The reasons for mental illness are a scientific question, and a diagnosis of 'evil spirits' rests instead on a claim of personal knowledge by witnesses. However convincing you may find someone's personal testimony, science isn't a private activity; it goes on in public and its claims are tested. If the results of those tests are to be accepted they need to be repeatable, not just by the person making the claim but by anyone.

Our understanding of the world is built up this way. Conditions that cause a disturbed person to thrash around wildly, speak in strange babbles of sound, throw off physical restraints and scream and curse may, to the devout religious imagination, look like demonic possession, but in some cases they're not even mental disorders. Instead they're symptoms of neurological diseases, such as Tourette's syndrome and – more commonly – epilepsy. There's a convincing case that such conditions as Tourette's and ergotism (the convulsive effects of ergot poisoning due to eating fungus-infected rye and other cereals) played a role in the entire witch craze of the sixteenth and seventeenth centuries.[28]

The scepticism that informs science is not averse to claims of reliable knowledge (for example, we know that demons and ghosts are mythical rather than real); it's averse, rather, to dogmatism. There are important areas of science where our knowledge is not only reliable but definitive. There may be theoretical developments in the laws of physics, but they supplement rather than overturn Newton's three laws encapsulating the behaviour of bodies in motion. There are also areas of science where our knowledge is definitive and complete. We know for a fact that all naturally occurring chemical elements have been discovered, just as all the continents on earth and the moons in orbit around it have been located. There aren't any more moons just waiting to be divined by a visionary astronomer through the power of intuition, let alone by a mystic receiving messages from the spirit world. We hope to find, and there is nothing to say we can't, a complete explanation for mental disorder too.

Depression and the mind

We are obliged to look for naturalistic explanations of depression. And though it remains a mysterious malady, we can at least be guided by rediscovering ancient wisdom. I mean this in the sense of the Hellenic tradition, not the biblical one: Hippocrates' thesis that disorder is due not to the introduction of spiritual influences from outside the body but to the workings of the brain within it. There are many oddities and competing explanations about how depression arises. But we do have a compelling explanation of how the mind works, even though it runs counter to most people's intuitions and thus typically goes unmentioned in public discussion of depression.

The explanation goes like this. All of our thoughts, feelings and experiences are physiological activities in brain (or neural) tissue. Mental states, abnormalities and disorders must come down to physiology, because that's all there is. To understand how we think, we need to understand how nerve cells behave (and especially how they interact with each other). The Nobel laureate Francis Crick, famous for his work on the molecular structure of DNA, described it this way: 'The astonishing hypothesis is that "you", your joys and your sorrows, your memories and your ambitions, your sense of personal identity, and free will, are in fact no more than the behaviour of a vast assembly of nerve cells and their associated molecules.'[29]

Crick's use of the phrase 'no more than' is deliberate. He's not just saying that the mind depends on the brain. Rather, the brain is all there is to it. It's a claim that biologists and psychologists refer to as reductionism, which means that a complex system can be explained by analysing its simplest physical elements and mechanisms. And while Crick calls it an astonishing hypothesis, that's not because it's an idea that scientists would regard as contentious or even novel. It's astonishing rather that something material and tangible – our bodies – can give rise to something so complex as an infinity of thoughts and emotions. And finally, the hypothesis is astonishing because it contradicts a common-sense belief (or perhaps a stubborn hope) that, because we are sentient beings, we must somehow be more than just the sum of our physical parts.[30]

Whence does this intuition derive? This is my guess. Humans maintain very hardy instincts about our place in the universe. We like to think that we are beings who choose our own destiny. We want to believe that we have free will and that consciousness is free-floating within our brains. If we are merely collections of machinery, this would seem to devalue us. If, alternatively, we have the ability to make unforced decisions, then we have grounds for hoping that, whatever mess we've made of life till now, we will choose well and responsibly in the future. We are moral as well as thinking beings, who have obligations to others. Many people, if pressed on the notion that they might not have free will, would regard the question as a clever piece of sophistry that offends against common sense. Thus does James Boswell recount the answer given by Samuel Johnson: 'Dr Johnson shunned to-night any discussion of the perplexed question of fate and free will, which I attempted to agitate. "Sir, (said he,) we *know* our will is free, and *there's* an end on't."'[31]

We can tell, at least indirectly, that Dr Johnson's common-sense response must be how most people think of our species. We just *know* our will is free, and there's an end to the discussion. Polling evidence suggests that, while belief in God has been declining in Britain since the mid 1990s, belief in an afterlife has remained stable in the same period. Somewhere between two-fifths to a half of all British adults believe there is a hereafter in some form. The notion that we survive death is comforting to many but it can't just be a wish, or it would be like the way adults profess belief in Father Christmas. It must be based on some idea of consciousness, however inchoate, that entails our mental states being separated from our bodies.[32]

Mind and body: the necessity of reductionism

René Descartes, commonly known as the father of modern philosophy, theorised that the mind and the body were two separate entities. This is what's known as dualism. As Steven Pinker, the cognitive scientist, puts it, Descartes 'became the laughingstock of scientists centuries after him (unfairly) because he proposed that mind and matter were different kinds of stuff that somehow interacted in a part of the brain called the pineal gland. The philosopher Gilbert Ryle ridiculed the general idea by calling it the Doctrine of the Ghost in the Machine (a phrase that was later co-opted for book titles by the writer Arthur Koestler and the psychologist Stephen Kosslyn and for an album title by the rock group The Police).'[33]

Yes, the mockery of Descartes was unfair. He didn't, contrary to popular belief, imagine that the essence of humanity was a pure mind that just happened to be transplanted into, and temporarily accommodated by, a body. Rather, he asserted that human nature could only be understood as a real (or, as he put it, substantial) union of mind and body.[34] Even so, this essential Cartesian idea has to be wrong. It still imagines that there are two substances, mind and matter, rather than one. We know this can't be true because of, among other things, the effect of anaesthetics, which are physical substances that induce a temporary loss of sensation or awareness. Ryle convincingly argued against the dogma of the ghost in the machine, even though this notion maintains a hold on popular understanding of what our minds consist in.[35]

What is this ghost in the machine imagined by the dualists? It is a non-material entity that temporarily inhabits and controls the body. And, like ghosts in haunted houses or on dark moors, it does not really exist. Reports of near-death visions or out-of-body experiences are sometimes cited in popular discussions as demonstrating survival after death and a separation of mind and body. They don't. These conditions are actually hallucinations caused by cerebral anoxia, where bodily injury or heart stoppage (even temporarily) deprives the brain of oxygen. The experiences of travel outside the body may be intense, and they are a genuinely interesting area of psychological and neuroscientific inquiry, but they are no more 'real' than were the drug-induced hallucinations of Thomas De Quincey in *Confessions of an English Opium Eater* that he was 'buried for a thousand years in stone coffins, with mummies and sphinxes, in narrow chambers at the heart of eternal pyramids. I was kissed, with cancerous kisses, by crocodiles; and laid, confounded with all unutterable slimy things, amongst reeds and Nilotic mud.'[36]

There's no reason to believe that consciousness exists, or possibly can exist, outside the body. We have no experience of any such state. We do, on the other hand, have much evidence to support Crick's astonishing hypothesis. Here is just some of it. Neuroscientists are increasingly able to record the electrophysiological signatures of thought and emotion. We know too that if a patient has their exposed brain stimulated during surgery they will have an experience that seems completely real to them but isn't. In a reported case of a woman suffering from epilepsy, a grid of electrodes was placed on the cortex (the outer surface) of her brain. These enabled researchers to record the electrical activity of her brain and also to stimulate different parts of it. If one area of the right hemisphere, called the angular gyrus, was stimulated, the patient had an out-of-body experience – not real, but so vivid as seeming to be real.[37]

A reductionist explanation of consciousness doesn't dispute that thought, emotion and perception are complex. But a reductionist approach is not a simple one. We know that the human brain is itself an almost inconceivably complex organ. The sheer scale of it is immense. I've cited the estimate that the brain contains around 100 billion neurons, or nerve cells, and these are connected by some 100 trillion synapses (tiny gaps between neurons), which pass electrical or chemical signals across neurons. This is how information travels between neurons in the mammalian brain (it's not just us). It's a chemical process involving what are called neurotransmitters. These are a sort of chemical messenger that transmits signals across synapses from one neuron to another. Within the nerve cell, the transmission is electrical. It's this electrical 'firing' of a neuron that releases the neurotransmitters to make contact with other neurons. The science of mapping the brain demonstrates that we can gain reliable knowledge of how it works, but also how limited still is our understanding of it.

I don't wish to appear dogmatic on this. There are a variety of different reductionisms in the philosophy of mind. It's one thing to say the mind is dependent on the brain; it's a plausible thesis that cognition depends also on the body, and on an interaction between the body and its environment (a position known as enactivism). Moreover, it's possible, as has been noted by the American psychiatrist Kenneth Kendler, for the science of the mind to be 'too biological' in its approach. It would be an error, for example, for psychiatry to focus at the level of subatomic particles in understanding mental illness.

Kendler argues for what he calls explanatory pluralism in place of biological reductionism. This involves a range of different and complementary perspectives in explaining mental processes and specifically psychiatric disorder. Looking at basic brain biology is not the most efficient explanatory level for explaining a psychiatric illness such as eating disorders among adolescent girls, for example, where culture is an important contributory factor.

An understanding of philosophy of mind matters for gauging an effective psychiatric response to mental illness. But my weak claim is that we don't need to resolve the philosophical question of the relation of mind and body in order to help sufferers from mental disorder. One philosopher, George Graham, has put forward what he terms metaphysical ecumenism, embracing even dualism, in arguing that 'our ability to explain and to understand a mental disorder has nothing directly or immediately to do with the truth or falsity of dualism and it is not threatened by the truth of physicalism . . .'. This seems to me eminently pragmatically sensible in getting depression better known and understood in public debate. And in principle depression probably requires understanding the mind at different levels of explanation and trying to see how they interact. Even so, my stronger claim is that there is a cost to regarding dualism as a possible model for the mind, for it may suggest that mental disorder is not truly an illness and that the fault lies with the sufferer. An artificial distinction between maladies of the mind and those of the body will tend to reinforce the stigma attaching to mental illness.[38]

Brain areas and what they do

There's a history, and not always a comforting one, to scientists' attempts to map cognitive functions to particular parts of the brain. Early efforts to do this were the purest superstition, and later ones – embraced by Nazi eugenicists – were founded on scientifically bogus and morally abhorrent notions of racial superiority. A now-defunct pseudoscience called phrenology gained currency in the first half of the nineteenth century by speculating that a person's psychological characteristics were determined by the proportions and size of their controlling organs within the brain. Hence (so it was argued) it was possible to determine someone's personality by measuring the shape of their head, because the part of the brain associated with a particular faculty would be more pronounced and bulge outwards.

The phrenological movement had immense popular support from fashionable opinion: Richard Whately, the Anglican Archbishop of Dublin, declared in 1836 that 'I am as certain that Phrenology is true as that the sun is now in the sky' and defended the notion against the charge that it was ungodly. It was all false – utterly false – because, among many other methodological failings, the faculties that phrenologists hypothesised had no empirical support and measuring the skull tells you nothing whatever about personality.[39]

Unusually among pseudoscientific myths, which have a tendency to hang around, phrenology disappeared completely. As far as I'm aware, it no longer has adherents and is merely a historical oddity. Numerous other weird nostrums have been given an afterlife in the digital age, including even the quintessential false belief that the earth is flat. Phrenology stubbornly refuses to be resurrected. Yet the idea of localisation, that parts of the brain serve specific functions, was not itself wrong. Partly to refute phrenology, a new discipline of brain physiology developed in the middle of the nineteenth century that genuinely did map functions to regions of the brain.

Paul Broca, a French surgeon, demonstrated in 1861 that a severe speech defect was linked to a particular part of the brain, in the left frontal lobe. A patient called Louis Victor Leborgne had entered a Paris hospital twenty-one years earlier having lost the power of speech. He could utter only a single syllable, 'tan' (or, when in an excited state, the same syllable repeated: 'tan-tan'). Yet his intelligence and cognitive faculties appeared to be unaffected, and he understood all that went on around him. Indeed, he appeared keen to talk but was unable to. On Leborgne's death, Broca conducted an autopsy and found a large lesion (damage to the brain caused by disease or injury) in a restricted part of the left frontal lobe (the posterior inferior frontal gyrus).

This was a single patient, but the finding mapped a function, the production of language, to an identifiable area of the brain. It was a groundbreaking discovery. Broca went on to examine eight further cases of similar speech impairment among patients who had suffered local damage to the same part of the brain. The impairment, where the patient has lost the power of articulated speech, is now called aphasia and the part of the brain where Broca located the damage is known as Broca's area. Further research a few years after Broca by Carl Wernicke, a German neurologist, identified another part of the brain, in the posterior portion of the left temporal lobe, as being associated with understanding language. Patients who had lesions in this area retained the power of speech but were unable to comprehend and hence speak coherently.

These observations by Broca and Wernicke have been independently confirmed since, many times over. Neuroscientists have identified a sort of two-lane neural loop running round the Sylvian fissure (a large diagonal cleft that separates the temporal lobe from the frontal and parietal lobes) in the left hemisphere of the brain.[40] This loop, made up of a large bundle of nerve fibres, connects the separable neural units responsible, respectively, for recognising speech and for producing it.

This is a simplified account of early explorations of the brain and scientists have made many important discoveries since, but these findings about language remain among the most important in identifying what makes us human. As one recent study on the causes of aphasia notes: 'From the earliest days of cognitive neuroscience, detailed studies of patients with acquired cognitive deficits following brain damage have provided the strongest evidence regarding the neural basis of cognition.'[41]

This is the astonishing hypothesis asserted by Crick: all our thoughts, experiences, emotions, perceptions and feelings come down to matter, to the material stuff that's inside our heads. I've given the example of language because it illustrates better than anything the remarkable capacity of the human brain. Language has two characteristics that set it apart. First, it's uniquely human. Only our species, of all that have ever lived, has developed the means to communicate by a linguistic system in which meaningful signs are made up from meaningless phonological segments, and can be combined to form complex signs which allow more or less any thought to be expressed. Second, language is a universal human faculty. It's not like other learnt behaviours, such as playing the violin or driving a car, which some people can do and others can't, depending on whether they've had explicit tuition. Language is acquired by everyone (excepting only those with the misfortune to have been born with brain damage or functional impairment) regardless of education, intelligence or any other factor that distinguishes humans one from another.

Other species may have sounds, but there is only one meaning that each sound can carry – a purr of contentment, say, or an urgent squawk to denote danger. Humans vocalise sounds and combine them to form words, phrases and clauses. There is no limit to the expressive power of this faculty of language. We can always add a clause to a sentence to expand it further, making the range of meanings infinite. Linguists refer to this combining of sounds to create meaning as 'the duality of patterning'. It works like this. Humans create units of language that have meaning (like words, which can themselves be separated into grammatical units known as morphemes) out of smaller units of sound (known as phonemes) that have no meaning by themselves. Single components of language may be known to other species, but the combination of grammatical units is an exclusively human activity. A parrot may be able to repeat words, but it doesn't have the mental capacity to put them together in such a way as to discuss ideas or hypotheticals or counterfactuals. No parrot will be able to talk about an event that might have happened three weeks ago if only something else had happened to cause it. Like all non-human organisms that have ever lived, parrots lack the faculty of language.

All of this comes from the brain, and from identifiable parts of the brain. Indeed, because all societies have language, some scientists conclude that the capacity for it stems from a unique property of the brain. They surmise that all neurologically typical babies are born with a unique language-acquisition device, or language organ. The effortlessness with which children acquire complex grammatical constructions by the age of three is powerful supporting evidence, though not conclusive proof. (A competing and widely held view is that language is the realisation not of an innate language faculty but of general-purpose learning mechanisms.)

Empirical support for the innateness theory, which is especially associated with Noam Chomsky, the theoretical linguist, and Steven Pinker, has been slower and more limited in coming than some of its advocates hoped half a century ago. But that's not relevant to my more fundamental point here, which is that the capacity for language, the most distinctive of human characteristics because we all have it and no other species does, can be mapped physically to matter. It is associated with distinct, identifiable areas of the human brain. Most theoretical linguists would regard language as a cognitive system in the mind of a speaker, and our cognitions are the outcome of the interaction of neurons and neural pathways. And we've got this way, with the distinctive characteristics that make up our species, just as every other organism has done, by the process of natural selection and random mutation, as theorised by Charles Darwin in *On the Origin of Species* in 1859.

The complexity of language is an example of the power of natural selection. And, though it's entirely a construct of the human mind, it has a material basis. This was discovered initially by the emerging science of neuropsychology. There are differing theories of how the brain organises language functions; an influential one developed in the 1960s by the American neurologist Norman Geschwind holds that each aspect of language (like understanding and speech) is served by different modules of the brain. We don't really know *how* the neural pathways of the brain work, but this is what produces our mental states. Crick's reductionist approach assumed that the different aspects of consciousness rely on a common mechanism. He sought to show this with the example of the neural mechanisms responsible for sight. Much work has been done in this field, and scientists have isolated some fifty distinct but interconnected areas of the primate visual system.

Early work in neuropsychology sought to identify particular lesions in the brain which might cause depression and other mental disorders. It was pioneering work, but it was constrained by two factors. Observable lesions are quite rare, and psychiatric disorders are quite common (so they can't be due just to damage to particular parts of the brain). In the late twentieth and twenty-first centuries, brain-scanning techniques have largely superseded this early science of neuropsychology. These enable neuroscientists to map mental functions to particular parts of the brain for healthy subjects as well as those with brain damage. They are a big advance in our understanding of the brain.

*

The development of magnetic resonance imaging (MRI) techniques has enabled scientists to explore the physical processes that take place within the brain when subjects report having mental experiences of things like sensations, perceptions or even emotions and desires. These processes are known as neural correlates of consciousness and the images suggest there is heightened activity in different parts of the brain during certain mental experiences. It's not that these areas work independently or sequentially to accomplish a task, but they have specialised functions and are interconnected with local areas that have other functions. It's like when we cross the road: we don't accomplish this task by, first, listening for traffic and only then looking for it before we conclude that it's safe to cross. In judging whether it's safe to cross the road, we perform these tasks – listening, looking, making a decision – simultaneously, with interconnected organs.

The theoretical work on human capabilities and the empirical findings on brain states are exciting, but they still don't reveal the origin of depression. MRI scans can show in great detail whether there are lesions in the brain of a depressed person, but they can't tell us even something so basic as whether the lesions cause depression. The question is not like Broca's area and the capacity for language. It may be that the depression *causes* the brain state, in a reaction to antidepressant medication, rather than the other way round. We don't understand enough at this stage. All we can say is that the answers to why humans get depressed do lie in the science of the brain. Even so, that knowledge tells us we're on the right track.

Mind and 'spirit'

There's one other reason, unconnected with the techniques of neuroscience, why we can determine with confidence that our conscious selves are purely the sum of physical processes and that there is no ghost in the machine. It's the absence of ghosts in the literal sense. Nobody has ever managed to show that it's possible to receive information by any means other than sensory perception, or to communicate with a spirit that has departed this life.

There have certainly been many attempts to communicate with the spirit world. Reputable and even great scientists have invested effort in this invariably fruitless venture. One sad case is Alfred Russel Wallace, the Victorian naturalist who independently of Darwin discovered that natural selection was the mechanism of evolution. Wallace turned to spiritualism in the 1860s, when he was in his forties. He maintained that the human mind was a new stage in the evolutionary process and believed that it was explained by non-material forces. In his view, the nature of the mind separated us from other species and meant that humanity was no longer subject to the operation of natural selection. His attempts to communicate with disembodied consciousness through spirit mediums were a testament to the ability of highly intelligent people to fall prey to charlatans and not be able to see through them even while knowing they sometimes cheated.[42]

Wallace was far from unique as a leading scientist giving credence to the notion of psychic communication. Sir Oliver Lodge, who in the late nineteenth and early twentieth centuries made important discoveries in electricity, radio and wave theory, was best known to the public as an advocate of spiritualism and a pioneer of psychical research. It was wishful thinking, driven by an urge to contact the spirit of his son, Raymond, who had been killed in action in the First World War. More recently, Brian Josephson, a Nobel laureate in physics, has speculated that developments in quantum theory 'may lead to an explanation of processes still not understood within conventional science such as telepathy, an area where Britain is at the forefront of research'.[43] In reality, there has never been a case under controlled conditions where consciousness has been shown to exist independent of the body. Scientists aren't adept at designing protocols in situations where the subject, in a test of remote viewing, extrasensory perception or communicating with the dead, is liable to cheat. For that you need someone who is expert at deception and creating illusions, namely a skilled magician. Nor is it credible to protest, as Wallace and his associates did and many have done since, that just because a purported psychic has been caught cheating once, it doesn't mean that all such effects and manifestations are suspect. Of course it does, because that's the most economical explanation for the effects.[44]

There is no sensible way of explaining how humans think and reason except by the physiological workings of the brain. We have no reason to doubt, and we have compelling reason to assume, that when these physiological activities cease, the party's over and the person no longer exists. The cadaver is not a bodily storehouse while the spirit marches on to paradise or perdition. *It is an ex-person.*

It's not easy to change habits of mind that suggest that dualism is true, that the mind and matter are different entities, and that just possibly we might live on even after electrical and chemical activity in the brain has ceased. We typically refer in everyday terms to someone of high attainment as having a very good brain, but it's actually a figure of speech. There is no separate person to whom the brain belongs, and no substance of serendipitous intelligence-stuff secreted within that brain. The philosopher Daniel Dennett gives as an example the 'genius of Bach [which] can likewise be disassembled into many acts of micro-genius, tiny mechanical transitions between brain states, generating and testing, discarding and revising, and testing again'.[45]

Depression and chemical imbalance

In short, our mental states are material and mental disorder must be at root physiological. There's nothing else they can be, because within the brain there's nothing else there. Neuroscience is able to say there are functions associated with different parts of the brain, and it can contingently identify them. A rough summary is that the frontal lobe is associated with reasoning, motor skills, cognition and producing language; the parietal lobe with processing sensory information (like pressure and pain); the temporal lobe with interpreting language and forming memories; and the occipital lobe with interpreting and processing visual information (like colours, shapes and written words).

Though there is no single cause of depression, or at least none that we can yet identify, there is a commonly cited explanation in public discussion. It is that depression arises from a chemical imbalance within the brain, specifically a deficiency of neurotransmitters. The theory about an imbalance is known as the monoamine theory of depression, referring to the group of neurotransmitters. It surmises that if levels of a particular chemical, especially one known as serotonin but also other neurotransmitters including norepinephrine and dopamine, decrease then this contributes to low mood and depression.

Serotonin is a neurotransmitter in the brain and other parts of the body, meaning it transmits messages between nerve cells, regulating their intensity. Scientists believe that it affects mood and happiness. Illicit drugs such as ecstasy produce higher quantities of serotonin. The most commonly prescribed class of antidepressant drugs are called SSRIs (selective serotonin reuptake inhibitor). These are designed to block the absorption of serotonin in the brain and thereby stabilise mood.

I discuss SSRIs and other drugs in more detail in Chapter 6. But here is a summary: antidepressants work. There is clinical evidence and good science behind them. Quite how they work remains unclear, though, and their clinical effectiveness doesn't necessarily mean that a chemical imbalance produces depression, let alone that it does so uniquely. It's the same problem as trying to work out whether depression is caused by small lesions in the brain or whether this damage is caused by the depression. Similarly, a deficiency of serotonin may be a symptom of depression rather than an explanation for why it happens.[46]

There is a stubborn statistical pitfall that researchers are required to navigate when interpreting data. The Canadian humourist Stephen Leacock (in his day job he was an economist) encapsulated it more than a century ago: 'A few years ago I went all round the British Empire delivering addresses on Imperial organization. When I state that these lectures were followed almost immediately by the Union of South Africa, the Banana Riots in Trinidad, and the Turco-Italian war, I think the reader can form some idea of their importance.'[47] To put it more succinctly but less evocatively: cause-and-effect relationships between variables are not always easy to infer.

Depression and neural circuitry

We can conclude that mental experiences derive from brain states, and neuroscience has made great advances in mapping them to particular areas of the brain. Yet the roots of mental disorder are not known. While depression has a physiological basis, it does not follow that it must be due to an abnormality of neurotransmitters. There are many variables and our knowledge of how they are interconnected is still quite basic.

The study of the brain is encompassed in neuroscience. Its investigations lead to the biological origins of thought and feeling. One theory is that depression is, and can only be, the outcome of disturbances of neural circuitry. Everyone has the same neural circuits. This is just part of the definition of our species, in the same way that everyone has the same pattern of DNA molecules to determine growth and development. The tuning of these circuits varies from person to person, however. The low moods that we all feel at times, whether or not there is an identifiable external trigger, are an outcome of brain chemistry.

It takes very little difference in the circuitry for one person to be locked into a cycle of sadness and clinical depression, whereas another recovers easily from temporary sadness and resumes a normal life. Studies of human behaviour show how normal reactions can feed on each other, producing a positive feedback loop (positive here does not mean something good, like the management jargon of 'positive feedback', but rather that the effects of a small disturbance are continually amplified).

This sort of process happens in complex systems. It may seem a forced analogy yet I recall thinking of it this way, when I tried to make sense of what I was going through (the mental world of depression) in terms of the subjects I knew as a journalist (the social world of economics and politics).

A quintessential complex social system is the economy: the interaction of numerous simultaneous decisions by consumers, businesses and investors, which are unpredictable even in principle. In the financial crash of 2007–09, the greatest economic disaster of the post-war era up till the coronavirus crisis, the banking system froze up completely. That system is the conduit by which credit is allocated to consumers and enterprises that need it. Banks typically borrow short-term, from other banks, and lend long-term, to businesses, homeowners and consumers. The difference in the interest paid and the interest received is known as the net interest spread; a high spread equates to a higher profit margin. But suddenly, banks that lent money to other banks worried they wouldn't get it back.

This produced a feedback loop. Banks would not lend, thereby causing other banks that depended on borrowing in the wholesale market to have a crisis of liquidity. Retail customers of these banks worried that they would lose their savings, so they queued round the block to get their money out. The banks, which hold in cash only a fraction of the savings they're responsible for, rapidly became insolvent. What was rational and prudent for each individual bank was disastrous for the financial system overall.

That's a widely accepted explanation of how an economic system of immense complexity, with numerous individual agents, came to a state of collapse.[48] I could imagine how, by analogy, a feedback loop might also happen with the brain. Particular episodes in life will provoke caution, sadness or fear. This is part of the cycle of apprehensions and emotions. The downward phase may, however, feed on itself, aggravating the mental disturbance. And like a complex system in the social world, the downturn may not be self-correcting but instead spiral downwards.

It's just a guess, though. It does not follow that mental disorder is the result of neurological disorder. And there may be a more powerful psychological correlate than a biological one. Depression often coincides with deep sadness at momentous events in life. Amid thwarted hopes, dashed illusions and the loss of those we love, we face only the weary inevitability of taking arms against a sea of troubles. It's not weakness to find such a prospect too much to rationally bear. The approach devised in the 1960s by Aaron Beck known as cognitive-behavioural therapy (CBT) stresses that clinical depression can be caused by distorted thinking. A stressful event, such as bereavement or the break-up of an important relationship, can stimulate a self-reinforcing chain of negative thoughts and stress. CBT works to correct these disorders of thought.

Depression is explicable in the terms of psychology as well as biology. Investigations of depression and effective remedies for it need to encompass both disciplines.

Depression and the immune system

As we have seen, dualism, the distinction between mind and matter, has a stubborn hold in our culture. This model of what we're like as humans is so ingrained that it's intrinsic even to the way health care treats the sick. Edward Bullmore, a Professor of Psychiatry at Cambridge University, has noted ruefully what he calls the medical apartheid of the disciplines of mental and physical health: 'These days, specialist medical services are routinely split down the middle, following the Cartesian divide between body and mind. Patients see a physician, who attends to the physical aspects of their disorder, or they see a psychiatrist or psychologist, who attends to the mental aspects. Physicians and psychiatrists are separately trained as specialists in one or the other of the dualist domains. Cross-talk is not encouraged.'[49]

The model of healthcare is false, because the distinction between ailments of the body and the mind is artificial. Bullmore proposes an explanation for depression, a quintessential mental disorder, that is 'physical' in origin. He proposes that levels of inflammatory proteins are a cause of depression. His argument goes like this.

The science of immunology explains how the body makes an inflammatory response (through microscopic workings of the immune system) to invasion by hostile bacteria. For example, if you're cut with a knife, then swarms of macrophages (specialised cells that detect and destroy harmful microorganisms) will overwhelm the bacteria. You'll notice the symptoms of acute inflammation after a day: redness, swelling and tenderness in the infected area. It would be completely consistent with what we know of the immune system that the same effect is visited on the brain, with depression as its outcome.[50] A further piece of evidence in support of this theory is that depression is to at least some extent hereditable, though the influence of specific genes is hard to pin down and is in any case tied up with environmental influences.

Depression's diverse explanations

There are hence many explanations for depression, and they are not limited to the ones I've mentioned. This isn't unusual with illness and it doesn't mean that the condition is imaginary. A disease that is directly observable like lung cancer can have diverse origins too, and so it does, though with that illness we know that the great majority of cases are linked to a single cause (smoking) and a link between that cause and the effect has been clinically demonstrated. With depression, by contrast, there is no study that definitively establishes a dominant cause and a predictable effect. The most that clinical studies have so far shown is associations between various changes in the brain and the incidence of depression. The prudent conclusion for now is that there is no single explanation for depression, which has a variety of interacting causes.

But uncertainty needn't entail bafflement. There are some explanations we can rule out and minimum criteria we can specify that accord with evidence. Depression is physiological in origin, not spiritual. Consciousness is the product of the brain, just as sight is the product of the eye. These are physical organs of immense complexity. Medical specialists understand completely how sight can deteriorate over time by clouding of the lens of the eye (cataracts), or damage to the optic nerve (glaucoma) or the retina (macular degeneration). The causes of mental disorder are much less clear than the causes of deteriorating eyesight, but they will eventually be uncovered by a fuller understanding of the workings of the brain. The artificial division between medical doctors of the mind and those of the body hasn't prevented important discoveries in neuroscience, but it's unlikely to be of help in enhancing public understanding of the nature of mental illness.

Beyond this conclusion we struggle for knowledge. Depression may be the outcome of the brain's chemistry or of genetic influences, and it may be driven too by psychological factors. And because the causes of depression are diverse, so may be the effective remedies. Some treatments will work for some sufferers but not for others. But their effectiveness doesn't depend on catching the depression early and in mild form. What matters is diagnosing the nature of the depressive disorder accurately. There is hope for sufferers of severe depression too. Intuitively you'd expect an illness to be more resistant to treatment the more entrenched and severe it is. In fact, the most common treatments, which are pharmacological and psychological, do work with severe depression whereas milder cases can be harder to shift. Having had severe depression diagnosed, I got intensive treatment and it proved not only effective for my illness but also invaluable for my quality of life beyond it.

3

HOW WE UNDERSTAND DEPRESSION

> To those of us who have suffered severe depression – myself included – this general unawareness of how relentlessly the disease can generate an urge to self-destruction seems widespread; the problem badly needs illumination.[1]
>
> William Styron, 1988

To be stricken with fear, anguish and revulsion at the self may have a proximate trigger but can't be explained by it. The relative weights of cause and effect don't balance, and the baldness of the irrationalism of my depression troubled me as much as the condition itself. Far greater minds than mine have referred to this imbalance of apparent cause and extremity of effect. As on much else concerning the human psyche, no writer in English has perceived depression more clearly and described it with greater acuteness than Shakespeare. When Lear is reunited with Cordelia and prepares for death, he is in a state of emotional torment:

> You do me wrong to take me out o' th' grave:
> Thou art a soul in bliss, but I am bound
> Upon a wheel of fire, that mine own tears
> Do scald as molten lead.[2]

The entire tone of *Hamlet* is set by the protagonist's melancholy caused, before the action starts, by his father's death and his mother's precipitate remarriage. There is a matchless description in Act 2, Scene 2, of the depressive's condition:

> I have of late, but wherefore I know not, lost all my mirth, forgone all custom of exercise; and indeed it goes so heavily with my disposition that this goodly frame, the earth, seems to me a sterile promontory, this most excellent canopy, the air, look you, this brave o'erhanging firmament, this majestical roof fretted with golden fire, why, it appears no other thing to me than a foul and pestilent congregation of vapours.

It's so evocative of what depression is like that many modern readers assume Shakespeare must have known the condition himself. That's an unwarranted inference, rather like assuming that because Shakespeare depicted the guilt of Macbeth with terrifying vividness, he must have known what it was like to be a murderer. It's safer just to accept that while we can't read the dramatist's biography from his works, he understood the human condition in myriad ways.[3]

The Edwardian critic A. C. Bradley saw depressive disorder as being integral to the puzzle of why Hamlet delays:

> [Melancholy] accounts for the main fact, Hamlet's inaction. For the *immediate* cause of that is simply that his habitual feeling is one of disgust at life and everything in it, himself included – a disgust which varies in intensity, rising at times into a longing for death, sinking often into weary apathy, but is never dispelled for more than brief intervals. Such a state of feeling is inevitably adverse to *any* kind of decided action; the body is inert, the mind indifferent or worse; its response is, 'it does not matter', 'it is not worth while', 'it is no good'.[4]

Everything here speaks of Hamlet's weariness born of what would then have been called melancholy. It is what a sensibility of the twentieth and twenty-first centuries would call depression, which is potentially the lot of every thinking and self-aware being and can strike anyone of any age.[5]

Depression is, I believe, the principal human pathology that can be better understood by integrating works of the imagination with the evidence-based methods of critical inquiry. But the task of understanding, both in the technical sense by scientists and in the emotional solidarity extended in public debate, still has far to go. And there are human reasons, often sympathetic ones, why we fail to inquire closely and find it easier to withhold curiosity. A harrowing case of self-destruction by an outstanding writer many years ago illustrates the extent of benevolent incomprehension by outsiders, which drove another noted author to anger. The first of these authors was Primo Levi, the Italian chemist, writer and Holocaust survivor, who took his own life in 1987 by leaping down the stairwell of his apartment block. The second was the novelist William Styron. Let me tell their story and explain Styron's protest.

'Depression is a disorder of mood, so mysteriously painful and elusive in the way it becomes known to the self – to the mediating intellect – as to verge close to being beyond description,' wrote Styron.[6] It was, he thus argued, almost incomprehensible to those who have not experienced it in its extreme form. That was in 1991. For all the discoveries in the science of the brain that have been made since, these have largely not entered public discourse; Styron's words might as well have been written yesterday. They might equally have been written centuries earlier. Public understanding of depression in a clinical sense, distinct from commonplace despondency, remains slight even after millennia.

I've stressed that depression is known and observed by its symptoms rather than its causes, and I've cited the medical listing of those symptoms. But the manuals and the clinical guidance are for medical practitioners. They can't fully convey what these symptoms mean to the sufferer. The symptoms typically include a lassitude and torpor that merge into despair. Writers and thinkers have dwelt in every age on the dread of death and the passage to oblivion, observing, in the words of Francis Bacon, that 'men fear death as children fear to go in the dark; and as that natural fear in children is increased with tales, so is the other'.[7] Depression is the dread not of leaving the world but of being in it, and of being seemingly alone in it.

The inevitability of extinction

Such fear will make little sense to those pundits who suspect that depression is a convenient clinical label for an inexplicable act of self-sacrifice or, less dramatically, an unwillingness to face the exigencies of life. And admittedly we all do have to face up, sooner or later, to a universal fate. There is inevitability in personal extinction and also in physical decline as we journey towards that state. It is a brute fact. Moreover, sadness in the loss of those we love, and anxiety for ourselves, is integral to human experience. The impermanence and brevity of our existence, and the universe's indifference to its end, are a tragedy born of our self-awareness. Arthur Schopenhauer, widely known as the philosopher of pessimism for his belief that the world is essentially irrational, concluded that the great insight yielded by religion is of our 'need for salvation from an existence given up to suffering and death, and its attainability through a denial of the will, hence by a decided opposition to nature'.[8]

But nature is nature, the human mind is a product of it, and mental disorders are real. Styron knew them in himself and observed them in others. His lament was elicited by Levi's death, which makes no sense outside the context of the depressive mind yet is completely explicable within it. Styron was shocked not only by the deed but by the popular response. Levi had emerged from the greatest atrocity of the modern age and been determined to describe its horror. His suicide several decades later seemed, to many critics, comprehensible only as an indication of weakness of character.

Some refused to believe his death was deliberate and preferred to think of it as an accident born of dizziness.

A typical quibbler, and a genuinely puzzled one, about the accepted account of Levi's death asked in a cultural journal the following rhetorical questions:

> Is it likely that a man who was a yea-sayer to life in his art would, like some quasi-Dostoyevskian character, hurl himself down a flight of stairs to his death? Is it likely that a writer with a profound sense of humor, curiosity, and a hunger for clarity and lucidity would quit life . . . ? Would a sensitive, devoted son 'desert' the ailing aged mother he was taking care of (as I have read somewhere)? Would a chemist kill himself by leaping down a flight of stairs? Would a sixty-nine-year-old man, forty-plus years out of Auschwitz, say 'I've had enough of life'?[9]

The implied answer to all these questions was no. The historically accurate answer to them has to be yes. According to contemporary newspaper accounts, Levi had not left home for several weeks preceding his death, his mother's illness was weighing on him, and he was troubled by new developments in historical scholarship suggesting that the Nazi depravities he had lived through and survived were far from being unique to the German experience. His mind may have been agitated by these and other factors. We don't know. Yet that is explanation enough, even so, for his disinclination to continue the habit of living. Even if none of these things had disturbed Levi, something else might have done. The threat of disturbance is always present. Mental turmoil is no less crushing in democratic domesticity than it is amid physical privation and oppression.

It's there, in his work. Levi wrote not only of the horrors of Nazism but also of their indelible residue. He observed: 'There exists a stereotyped picture, proposed innumerable times, consecrated by literature and poetry and picked up by the cinema: at the end of the storm, when "the quiet after the storm" arrives, all hearts rejoice.'[10]

All hearts rejoiced at the defeat of Nazi barbarism. But nothing was the same again. The quiet was not the reassertion of the ideals of European civilisation. It was the trauma of discovering the camps and the mass graves. Levi was one among the many millions of victims of Nazi oppression. Another, the Austrian novelist Stefan Zweig, had been in the 1920s the most popular author in Europe. He exemplified the learning of the central European polymath and the ethical humanism of European Jewry. He fled Austria in 1934 and found eventual sanctuary in exile in Brazil, but he never found rest. In 1942, when he was aged sixty, Zweig and his wife, Lotte, killed themselves with poison. In a suicide note Zweig said, poignantly, that 'the world of my own language sank and was lost to me and my spiritual homeland, Europe, destroyed itself'.[11] Nazism was by that time fated to wartime defeat, yet it was in one sense already victorious. It had destroyed a civilisation. It eventually destroyed the brave, brilliant figure of Levi too.

Styron was incensed at the incredulity and incomprehension displayed in popular coverage of Levi's tragic end. Commentators compared the agonies of the death camps with Levi's burden of recollecting them, and implicitly, even unwittingly, concluded that the psychological condition of depression was less painful than the experience of genocidal oppression and the continuous risk of entering the gas chambers. These mental states were, in reality, the same experience, from which escape proved impossible. The respective sufferings were incommensurable: they did not feature on the same scale of measurement. To die was not an irrational, still less an inexplicable, resort. It was an easier fate to bear and a swifter decision to discharge than to persist with the anguish of living.

Depression for Levi was finally insupportable. But the illness has many faces, some of which are easy for outsiders to overlook. For some sufferers, depression is a constant demoralising presence, like a tinnitus of the mind, that wears them down but doesn't destroy them. For others, it is a terrifying precipitous descent. What all variants and pitches of depression have in common is that, unlike a broken limb, their effects are observable to others only indirectly. The common experience of depression is solitude and despair. That sense of being alone isn't a conceptual error but a fact. The stigma of mental illness remains a powerful deterrent to admitting it, even to yourself. Hence the sufferer is doubly stricken, by disorder and shame. And because the pain of severe depression is all but impossible to hide in close quarters, those who care for a victim of it will suffer pain too. They will be perplexed, anxious and sometimes angered by the inability of a loved one to recover their senses, as if wellness were merely a matter of will.

Nerves, melancholia and the terminology of depression

The accounts of writers concerning the depressive state should impress upon us that to list symptoms, while necessary, is only a partial description. Depression is a term with both a technical and a popular meaning. In everyday language, to be depressed is to feel sad. It's natural to assume from this that to be clinically depressed is to be *very* sad. But depression in a clinical sense is not this. It is not even just feeling *very sad for a long time*, though that may be part of it. It is what clinical psychologists and psychiatrists call an affective or mood disorder. I've stressed that sadness is a normal part of the human condition but an affective disorder is characterised by *abnormal* emotional states. It comprises a host of symptoms that will differ from person to person. Some sufferers will be worn down by ennui – a sense of listlessness and hopelessness – rather than more specifically sadness. What formerly gave life a purpose no longer matters. What once provided pleasure palls. The appetite for food, recreation, conversation and human intimacy has, the sufferer finds, abruptly departed.

Physical diseases are recognisable. Mental disorders are less easy to pin down and differentiate. The psychiatrist Tom Burns is another critic of the expansiveness of 'depression' as a diagnosis. He describes the term as 'a catch-all for all neurotic psychiatric disorders, displacing older terms such as "nervous breakdown" or "nerves". Its advantage is that it is so familiar and non-stigmatizing. Its disadvantage is that it is just too all-inclusive. We have all been depressed but, luckily, most of us have not "suffered from depression".'[12]

The reference to 'nerves' underlines his point. Though there's been a recent attempt to revive this older term to denote depression, it hasn't taken hold. Edward Shorter, a historian of medicine and psychiatry at Toronto University, argues that 'nerves' (in the sense of a nervous breakdown, though doctors do not any longer use this term) is a more accurate label than depression in many cases. Depression has certain identifiable features, which Shorter considers biological, but the diagnosis of it has risen sharply in the era of psychoanalysis and especially, as we've seen, since the American Psychiatric Association's *DSM* a generation ago.

Shorter argues that the term 'depression' is a recent innovation and that it's confusing. He says that it was 'created' (his word) as a compromise between competing explanations for mental disorder. Psychoanalysts looked into a person's memories, thoughts, feelings and desires. Psychiatrists were increasingly concerned instead with biological explanations. These approaches are not the same. Indeed, they conflict. In order to bridge that divide, psychoanalysts and psychiatrists settled on using the term 'depression', each having their own interpretation of it. Yet the label ends up obscuring the differences between these approaches. It also lumps different illnesses into the same category of depressive disorder. In Shorter's view, there is a clear distinction between melancholic illness and non-melancholia. He defines these conditions this way:

> Melancholia, a grave form of depression involving slowed thought and movement, a complete joylessness in life and lack

> of hope for the future, had always been considered a separate illness. By 1980 the term melancholia had gone out of style and had been replaced by endogenous depression. The other form of depressive illness that psychiatry had always recognized as separate was an ill-defined aggregation of symptoms of mood, anxiety, fatigue, somatic complaints – and a tendency to obsess about it all – that had been called on occasion neurasthenia, neurotic depression, reactive depression and other terms indicating real illness but not melancholic disease.[13]

Hence, both these authorities, Burns and Shorter, regard the term 'depression' as unhelpful because it is too broad. It covers different conditions. Shorter argues for a return to older terminology: melancholia and nerves. This reversion to the vocabulary of an earlier age shows no sign of happening and there is a cultural reason why, as an illness rather than a technical term for a collection of neurons, 'nerves' is unlikely to be adopted. It retains its hold in the public imagination principally as the complaint of Mrs Bennet in Jane Austen's *Pride and Prejudice*: 'You take delight in vexing me. You have no compassion for my poor nerves.' Despite the heartlessness of deriding a foolish person, the reader can't help but take sly pleasure in Mr Bennet's mockery of his wife's fastidiousness. Nerves, so Austen implies with skilful artifice, are a rhetorical expedient with which to cloak a lack of verbal dexterity and capability in dealing with the social world.

Depression does admittedly have potentially confusing connotations, illustrated by Shorter's reference to it being 'endogenous'. Something is said to be endogenous if it has an internal cause or origin. Shorter is using the term carefully to indicate that depression, in the clinical sense, has a physiological origin. To many people with depression, though, their condition has been triggered by traumatic events in their personal, professional or public lives. This is consistent with the notion of endogenous depression but it is also a recipe for confusion in the public mind. You can imagine a patient responding with perplexity to a diagnosis of endogenous illness. They'd want to know what the term means, and on being told that their depression has an internal cause, they might be incensed at the unintended insinuation that they're just imagining it. Terrible things can trigger depression, but that doesn't contradict a diagnosis that the depression is internal.

So, for the moment at least, we're almost certainly stuck with the term 'depression', but we need to be aware of how variable it is. GPs frequently encounter patients who report feeling depressed and who may have some symptoms consistent with clinical depression. The doctors may refer these patients to psychologists or psychiatrists who have training and experience in distinguishing between low moods and mental disorder. But the boundary is indistinct. As Burns puts it: 'Psychiatrists try to identify different types of depression because there is such variation in how it responds to treatment. The hope is that if we can define the types that respond well to treatment we can make sure we identify and treat them early.'[14]

The strangeness of depressive disorder

The caution in that statement is salutary. It is a hope, not a confident appraisal, that clinical depression can even be identified, let alone classified by type. So let's look at this broader question of what depression consists of. Psychologists and psychiatrists categorise depression as having four broad characteristics: affect, cognitive strangeness, behavioural symptoms, and symptoms of external appearance.

From the standpoint of the sufferer, affect – or mood – is the prism through which the world appears. The mood of the depressed person may comprise despair, agony or prolonged listlessness. Oliver Sacks, the neurologist, wrote: 'We have, each of us, a life-story, an inner narrative – whose continuity, whose sense, *is* our lives. It might be said that each of us constructs and lives, a "narrative", and that this narrative *is* us, our identities.'[15] Depression breaks this sense of continuity. We see the world anew, but not in a revivifying sense. The surroundings are forbidding and strange, and this alien quality is baffling to a mind that previously had a sense of order and self-control.

The break in our sense of narrative leads to the second feature of depression: its cognitive strangeness. Sufferers from depression are apt to be self-critical, but not in the sense of striving to improve. Rather, the self-criticism manifests itself in ways so extreme as to defy rational explanation. Sufferers from depression are not just attuned in a normal and healthy way to their own imperfections, but instead magnify their weaknesses, real and imagined, in recalling significant recent events in their lives. For example, a common response to the break-up of a relationship is to feel regret, loss and loneliness. Few can feel elation even at the end of an unsuitable relationship; at most, there will be a sense of relief at what has passed but sadness that it was ever experienced at all. To those who suffer clinical depression, these normal human feelings are turned to shame and guilt. When friends rally round and offer support, your instinct tells you that it's a burden to them, that they're being merely diplomatic, and that you should stay out of their way.

Third, there are behavioural symptoms. These are not an act: when you're depressed, your movements become sluggish and your speech tails off into a monotone. You stare into space and speak only when spoken to. You dwindle into inconspicuousness, but with one exception. The external aspect of your personality that shows unfailing agitation is a constant susceptibility to crying, often with no apparent occasion and sometimes for the most ostensibly trivial of causes. You don't go out, and may stay in bed for days. Alternatively, some sufferers become highly animated and fidget, or show obsessive symptoms like pacing up and down or continually sighing.

Finally, there are symptoms of external appearance, sometimes triggered by these alterations in behaviour. Rapid weight loss caused by a drastic diminution of appetite is common, though not universal. For a few sufferers, depression manifests itself instead in weight gain, due not to expanded appetite but to the absence of human company. The immediate solace for them is to engage in eating, as this is at least an activity. For many if not most sufferers, the most visible symptom of depression is a drawn, pallid and fatigued appearance due to chronic lack of sleep.[16]

Seeking to recover the 'real' self

That's how it was with me. The dogma of the ghost in the machine retains a powerful hold on the human imagination, as we instinctively think of ourselves as more than the sum of chemical and electrical reactions. Even knowing its errors, I couldn't quite dispense with the superstition. It seemed intuitively obvious that there had to be a real me, a mind with its memories and recognitions that made me at home in the world, and that it had departed my mortal frame. It wasn't that I had taken leave of my senses but that my senses had taken leave of me. I wanted nothing more than for these senses, the 'real' me, to come back.

What filled the vacuum was a state unlike anything else I'd experienced. On the day that I lost recollection of my address, I eventually made it home by getting on a bus with a familiar number. I then wrote to friends with a vague intimation that I was unwell, crawled into bed and put the covers over my head. I didn't move for twenty-four hours. I told myself I had stress but had no conception of what that might mean beyond an accumulation of anxieties and disappointments that had taken over my life. I would imagine this 'stress' in the form of a cloud, outside the window of my bedroom and enveloping the street and the neighbourhood. It expanded and expanded. Mental illness was unknown to me. Though there is evidence of some genetic influence on depression I had no history of the illness, no knowledge of any family susceptibility to it, and, again, no real notion even of what it meant. It felt not only agonising but confusing. I wanted to know; I wanted to understand.

To me, the ideal in life had always been the determined application of reason. It might not bring happiness, but that was not its justification. It was the route to reliable knowledge, and to gain this was an end in itself and would be its own reward. The stoicism I'd sought to practise in my personal life was not an affectation but a consciously sought objective. When I was in my twenties I read extensively in the works of Samuel Johnson and found there, so I thought, some of the answers to the natural questioning we have at an early stage of life about its ends. These answers were not Johnson's own, for he trusted to providence and divine mercy, but they preoccupied him. And they impressed me.

The encounter of Prince Rasselas with 'the wise and happy man' in Johnson's only novel, *The History of Rasselas, Prince of Abissinia*, testifies to the Stoic vision: 'He enumerated many examples of heroes immovable by pain or pleasure, who looked with indifference on those modes or accidents to which the vulgar give the names of good and evil. He exhorted his hearers to lay aside their prejudices, and arm themselves against the shafts of malice or misfortune, by invulnerable patience; concluding, that this state only was happiness, and that this happiness was in every one's power.'[17]

It's satirical, of course. The novel is a brief fable with more than a passing (but entirely coincidental) resemblance to Voltaire's *Candide*, which was published in the same year of 1759. Both novels deal with the futility of the search for happiness. Both have spoken to me over my adult lifetime, and increasingly at its midpoint. Yet Johnson's, while less well known, is much the more profound of these works of the imagination. In his telling, the wise and happy man suffers (only a few paragraphs further in the narrative) the death of his daughter, and laments: 'My views, my purposes, my hopes are at an end: I am now a lonely being disunited from society.'[18] The comforts of reason count for nothing any longer.

But those comforts mattered to me and they also did, in a way, for Johnson even while he disclaimed them. In his epistolary *Rambler* essays of 1750–52, he disputes that philosophy can reconcile us to misfortune. Even so, 'the antidotes with which philosophy has medicated the cup of life, though they cannot give it salubrity and sweetness, have at least allayed its bitterness, and contempered its malignity'.[19]

It's a poetic way of saying that happiness is not within reach through the application of reason, but that tranquillity may be. This was a maxim I clung to through years of both advance and adversity. As disappointments and loneliness came to dominate my days, I thought of it still more. Merely succumbing to life's vagaries is pointless, and it wasn't my choice. The application of reason doesn't require quietism, and if things are wrong then the first duty of the rational being is to strive to put them right. But I held to the notion that a tranquil mind requires at all times reserve and dignity. And the experience of depression puzzled me even before it alarmed me, because reserve was no longer of any help. Instead of calming the fevered mind, the attempted exercise of reason only made me more agitated. Over that twenty-four hours I lay paralysed with misery, accompanied by groans and sobs. I wanted to scream but found that I lacked the energy even to do that.

This was the point in my experience where the common-sense notion of depression diverged radically from my experience of it. I wasn't just sad, but subject to a qualitatively different form of mood. The cloud of my imagination, which expanded across the rooftops and swallowed all beneath it, separated the world from where I lay. After that day, my mental world collapsed. Every sensory perception brought darkness. I didn't feel merely low or melancholy: it was a constant torment to be awake at all. I was overwhelmed with feelings of guilt, shame and worthlessness, amid a searing anguish that never lifted. I was haunted by the conviction that I was evil and, like a religious penitent, I yearned to confess and receive absolution. Yet it wasn't available, even in principle, for there was no one who could bestow it, and I was in any event unable to articulate the reasons for guilt or whence it derived.

Bowed by guilt, I would imagine hearing mocking, derisive voices. It was an episode where, to this day, I don't know for certain if my depression was psychotic. Technically, it can't have been. The voices didn't exist and I knew in principle that this was so, but I couldn't distinguish whether I was imagining the voices vividly or really 'hearing' them. At its extreme, this was the condition evocatively described in Evelyn Waugh's novel *The Ordeal of Gilbert Pinfold*, in which the protagonist, apparently overhearing (never directly witnessing) the scorn of fellow passengers on a cruise ship, recalls the lament of Lear: 'Now he was struck with real fear, something totally different from the superficial alarms he had once or twice known in moments of danger, something he had quite often read about and dismissed as over-writing. He was possessed from outside himself with atavistic panic. "O let me not be mad, not mad, sweet heaven," he cried.'[20]

The simplest pleasures, such as reading a book or listening to music, were by then far beyond me. I lost my appetite for any sensual experience, whether through sound or sight or taste or touch, or the habits of thought. The simplest tasks, such as opening the front door, became near-insurmountable challenges. All I could think of, on and on, was the oblivion of sleep, which wouldn't come, and how merciful it would be not to wake up again.

I've used the term 'chaos' to describe this state of deep depression, but that is a metaphor. In fact, depression, at least as I experienced it, was a stable state. Mine was immovable and impermeable. It was always there. Nothing could shift this monolith. I would typically have a moment's respite on waking, almost always after a troubled night, and then the darkness would envelop me once more. Merely the least of its symptoms was lassitude, which ensured that the condition was very hard to confront, let alone dispel. And the greatest symptom was terror, which encompassed almost any experience I was likely to have. I was terrified of being amid crowds and terrified of being left alone, terrified of the hubbub of daily life and terrified of the bleakness of introspection.

The limits of living by reason

In the voyage of life, I had overlooked an essential truth. Reason is what we aspire to, but it is not the natural state of us poor humans, at once capable of grasping the vastness of the universe and awed by our relative insignificance within it. I'd thought it was. It was literally part of my upbringing.

In the last year of her life, my grandmother resigned from the Liberal Jewish Synagogue in London because, as an atheist, she thought it the intellectually honest course. She told me about it matter-of-factly one day when I called on her for a cup of tea. Not having what many would regard as the consolations of faith, I found it admirable that she was honest with herself that reason was the light she chose to travel by, even as she faced the imminent end of her days.

I've always remembered it and admired the ability to conduct life in defiance of religious authority. Hers was a sundering of a relationship with the synagogue, in a far less dramatic way, that I imagined comparable to that of Benedict (or Baruch) Spinoza, the seventeenth-century Dutch thinker. Spinoza was excommunicated by Amsterdam's Portuguese-Jewish community. Why this happened and the reasons for the virulence of the ostracism, proclaimed in front of the

Ark of the Torah, are unclear and are matters for historical debate.[21] But I wanted that unflinching commitment too. I had read a little of Spinoza[22] and dwelt on what I fondly imagined was my own responsibility. It was this: 'Since the intellect is the better part of us, it is certain that if we want to really seek our advantage, we should strive above all to perfect it as much as we can. For our greatest good must consist in the perfection of the intellect.'[23]

When I lost hold of my mind and was plunged into turmoil, I found how hard it is to live by the dictates of reason alone. It wasn't the guide I'd taken it for. It wasn't a guide at all, for it yielded disastrous conclusions. Thinking myself capable of reflection and sound judgement, I'd acted on reason and, so it seemed, reason had led me into pathways of destruction of things I held dear. And because I had made those rational choices, I became beset by guilt and shame.

In fact Spinoza knew well the limits of rationality too, for, as he wrote, 'men are not conditioned to live by reason alone, but by instinct, so that they are no more bound to live by the dictates of an enlightened mind . . . than a cat is bound to live by the laws of nature of a lion'.[24] This is the lot of humanity. The flowering of reason is historically bound; the abandonment of reason is a threat that's ever present.

In my own state of confusion and despair, instinct was all I had left. Continuing to operate each day by habit and through instinct was the way I tried to cope with each trivial decision from hour to hour and day to day. After a disturbed night, it was an effort of will to face the morning and to slouch out of bed. From a pale and raddled reflection in the bathroom mirror, I'd know by instinct that I'd feel physical discomfort if I didn't shave and be ostentatiously unsightly when I was desperate not to be seen at all. By instinct, born of recollections of schooldays decades earlier and of maternal solicitude, I'd tell myself I had to eat breakfast, as it would be far too much of an effort to eat in the distant horizon of lunchtime. And so it went on through the day, each day.

Perhaps that would have been enough on its own, but it didn't seem that way. Some people operate with low-level depression, anxiety and disorder over years; some do so seasonally. They live with it and manage it because there is no other course available to them. It's like living with a phobia. If you suffer from dread of spiders or moths, or fear of flying or of public speaking, it's a debilitating mental disorder, but not one that overturns your life. You can try and plan to minimise your chance of encountering the object or experience that stimulates the terror. Nor is the fear necessarily irrational. There are sound evolutionary reasons why we feel a dizzying apprehension when we traverse a great height and then look down. The remedy (not the cure) for vertigo is not to go high. The trauma is containable, and contained, by not being given cause to erupt.

Depression and evolution

Clinical depression isn't remediable in this way; nor can it be compartmentalised, let alone shut out. It's less easy to see how depression could serve an adaptive function in the same way as vertigo, but it's possible. Consider anxiety. This is a mental state in which sufferers experience fear and apprehension, and have physical symptoms such as excessive sweating and palpitations. Anxiety has an adaptive benefit in making humans take care at times of danger. A generalised anxiety disorder means that the mental reaction doesn't return to a more normal state when the danger has passed. It's distressing and maladaptive, but you can see how anxiety in a more controlled way has benefits.

The persistence of low moods and their mutation into clinical depression may have adaptive benefits too. If the pursuit of dominance in a highly competitive pack leads to depression, then the loss of this potential source of conflict might stabilise the wider group. It could also prevent the longer-term disappointment of continued failure to achieve unattainable goals. This sort of effect has been observed in other mammalian species, as noted by George Brown, an academic specialising in the social settings for psychiatric disorder:

> The idea that depression is linked in some way to behavioral patterns in group-living mammals evolved to deal with conflict with conspecifics is persuasive. Experimental work has documented apparently depressive-like states in such animals after defeat, for example, in terms of almost immediately lowered testosterone, raised cortisol, and 'retardation' after a dominant male marsupial sugar glider is transferred to another group where his dominant status is lost.[25]

Brown argues that it's possible that some common forms of depression are linked to traits that have evolved to deal with conflicts due to loss of status; but he expresses doubts that severe depressive disorders have ever been of adaptive benefit. That makes intuitive sense, but we can't know. We don't have the evidence of mental states in prehistory that we do of the evolution of physical organisms, so we can only guess. We know (and I mean 'know' in the sense that we know any scientific proposition) that biological systems develop, as Charles Darwin said they did, through the mechanisms of natural selection and random mutation. Mountains of evidence confirm this finding. Nothing undermines it. But that doesn't mean (in a popular misunderstanding) that phylogenetically later organisms are 'better' than their evolutionary predecessors.

An example of a successful organism is the earliest, namely bacteria. Bacteria are more numerous and have greater biomass than any other organism, but they are simple. That indeed is one of the reasons for their success, along with their small size. They reproduce and adapt rapidly. Natural selection isn't a blueprint for progressive improvements: it explains how organisms adapt to existing conditions. As one of the great figures in evolutionary theory, Ernst Mayr, explained it: 'For instance, animals and plants in New Zealand had been selected to be adapted to each other. When English animals and plants were introduced to New Zealand, many of the native species, not being "perfect", that is, not being adapted to the invaders, became extinct. The human species is highly successful even though it has not yet completed the transition from quadrupedal to bipedal life in all of its structures. In that sense it is not perfect.'[26]

Mayr's reference to the uncompleted transition to bipedalism is a pregnant example. Humans evolved to walk on two limbs rather than four, with an erect posture, only recently in our ancestral hominin history, probably between four and six million years ago. Acquiring this skill requires changes in anatomy and especially the skeletal structure. The spines of our mammalian ancestors were originally arch-shaped. Walking upright caused our backbone to become a straighter pillar, bearing weight directly. This had adaptive advantages but also costs. Bad backs and fallen arches are the outcome of this evolutionary adaptation (which no intelligent creator would have designed). A much more efficient structure would be a torso that tilts forward, relieving pressure on the vertebrae, which in turn would require an upwardly curved neck with enlarged vertebrae to enable the head to stay upright.[27]

Depression may similarly have had an adaptive effect that is incompletely effective and causes problems. Perhaps it originated in feelings of sadness and disappointment and turned into a more severe disorder. But we don't know. All we have are speculative hypotheses. We can't know even if certain behaviours are direct adaptations or, like creating fire or inventing the wheel, merely applications of the highly complex organ of the human brain. We can formulate theories of how, say, altruism may have served an adaptive purpose, but we're unable to trace its evolutionary path. The theories may make intuitive sense. But they may just be fables that we tell ourselves, given the lack of any scientific justification.

The power of natural selection

The best we can do, and it is important even so, is to understand the power of natural selection to accomplish what superficially appear to be impossibly intricate outcomes. Yet people, and even leading scholars, often overlook this. The ability of humans to convey an essentially unlimited range of messages through language, again, is an ideal example because it is unique to our species. The great variety of human languages (there are about 7,000 natural languages spoken in the world today) is testament to evolution: these have evolved from earlier varieties, and palaeolinguists trace modern languages back to extinct forms. In a recent study, a team of linguists and biologists at the University of Pennsylvania examined a huge collection of English texts, in machine-readable format, from the twelfth to the twenty-first centuries. They wanted to find out whether language changes by chance or by some selective force. They found that language can change by random chance (though some kinds of change, notably simplification and the elimination of various irregularities, do recur in the history of language). It's very similar to the way physical organisms evolve by random mutation and natural selection.[28]

The parallels between evolution in the natural world and the evolution of language are close. Charles Darwin himself remarked on them. Yet there's an ineradicable problem for historical linguists in reconstructing the evolution of language. Unlike dinosaurs or trilobites, words don't fossilise. The evidence is necessarily indirect. One of the most gifted scholars Britain has ever produced made a historic contribution in helping uncover that evidence. This was Sir William Jones, an eighteenth-century philologist and scholar of India. While serving as a judge in Calcutta he learnt ancient Sanskrit, and was probably the first westerner to do so. He was astonished to find huge overlaps of vocabulary between Sanskrit, Latin and Greek – more, he wrote, 'than could possibly have been produced by accident'. He surmised that these languages must have 'some common source, which perhaps no longer exists'.[29]

It was brilliant reasoning. Just as humans and great apes have a common ancestor that no longer exists, so do these ancient languages. Scholars refer to this single source, dating back many thousands of years, as Proto-Indo-European. It gave rise to most of the European languages that are spoken today and many Asian ones. Before Jones, scholars wondered which was the first language, and generally assumed it to have been Hebrew. They were wrong but, before Jones's conclusions, they didn't have much to go on. There are people today who reject Darwin's theories because these contradict the testimony of scripture, and it's worth stressing that their aversion must be to the whole of scientific inquiry and not only to evolutionary biology. After all, the science of modern linguistics demonstrates that the story of the Tower of Babel recounted in Genesis, which to the ancients explained the diversity of languages, is mythical, just as astronomy, biology, palaeontology and geology demonstrate that the biblical account of creation is a human invention.

The evolution of language, in the sense of continual change in the way that speakers of any natural language use it, is magnificent and complex. And, like the evolution of hominids and every other living thing, it's unplanned. No one is in charge. Yet language never breaks down into chaos. On the contrary, every natural language, from the Australian aboriginal Warlpiri, spoken by around 3,000 people, to the far more widespread language that you're reading now, has an immensely complex system of grammar and can express a full range of meanings.

The power of natural selection as an explanation for the human condition, as well as the natural world, is immense and revelatory. But it doesn't produce 'perfect' outcomes.[30] On the contrary, as Jerry Coyne, the evolutionary biologist, has written: 'Although organisms appear well designed to fit their natural environments, the idea of *perfect* design is an illusion. Every species is imperfect in many ways. Kiwis have useless wings, whales have a vestigial pelvis, and our appendix is a nefarious organ.'[31]

I've concentrated on language as an example of how a complex system can be explained by Darwin. The same is true of literally any and every part of the natural world. In Chapter 2, I referred to Edward Bullmore's suggestion that levels of inflammatory proteins in the brain cause depression. He advances this theory with reference to the mechanisms of evolution that Darwin explained: 'When it comes to biological systems – or life as we know it scientifically – the answer to the question "why?" is always the same: natural selection . . . The ultimate reason for any biological phenomenon or phenotype to appear in life, or to disappear into the fossil record, is that it is more or less adaptive, more or less likely to make an organism fit to survive.'[32]

Why, then, has clinical depression persisted? Bullmore notes that there is on the face of it no survival advantage to having severe mental illness. On the contrary, on average it dramatically foreshortens lives. Yet there is increasing evidence that inflammation can cause changes in brain states and behaviour. Inflammation is the defensive response of the immune system to infection. Bullmore's hypothesis is that depression has its roots in the body's immune system: it may be (not in all cases but in some) an inflammatory response to pain.

This isn't a metaphor for the pain of mental illness. He means that physical inflammation can directly cause mental disorder and other psychological states. The dualist division between mind and body is not sustainable. Among the challenges it faces is the emerging science of neuroimmunology, which studies the links between the nervous system and the immune system. Bullmore foresees the development of new treatments for depression that work by reducing inflammation. It's an alluring prospect, and again it would be a consequence of greater understanding of the human condition owing to scientific inquiry. It's not an imminent advance or a universal cure but a promising intervention.

As Bullmore has said elsewhere: 'This won't be a panacea. I don't think inflammation causes all depression or that anti-inflammatory drugs will work for everyone with depression. That "one size fits all" approach is a limitation of where we are at the moment. The future, I hope, will take a more personalised approach, using anti-inflammatory interventions more precisely to target inflammation in those patients where it is likely to be a causal factor.'[33]

Perhaps in prehistory, genes were selected to increase the inflammatory response to infection in a way that conferred an advantage. We don't know quite how, because other people's thoughts and moods aren't open to external inspection even in real time, let alone in history or prehistory. All we can say is that, like the imperfections of our bodies or the idiosyncrasies of language, clinical depression is just the way things turned out for us humans. There's always a danger, in the absence of physical evidence, of alighting on an explanation that fits the available facts but is entirely made up. Scientists refer to these as just-so stories, akin to Rudyard Kipling's explanation, among many others, that the camel has a hump owing to a djinn's punishment for refusing to work.

We can speculate that depression had an adaptive purpose for our early hominid and even pre-hominid ancestors, by preventing them individually from persisting with ambitions that were bound to be thwarted and by collectively calming their intra-group disputes. Perhaps depression itself evolved into different forms, so that the widespread phenomenon of seasonal affective disorder (SAD), which affects women in particular, derives from Ice Age times and the biological tendency to slow down in winter.[34]

It's a theory, but is it true? Don't ask *me*. We can never know, even in principle, because the data isn't there and can't be recovered. It died with those who suffered these prehistoric instances of depression. We have suggestive evidence from the practice of prehistoric surgery and the behaviour of non-human species. But only in the era of written testimony of depressive disorders, beginning roughly 2,500 years ago, do we get some idea of what it was like then and compare it with the experiences of sufferers now.

4

HOW WE MISUNDERSTAND DEPRESSION

> Great wits are sure to madness near allied,
> And thin partitions do their bounds divide.
>
> John Dryden, 'Absalom and Achitophel' (1681),
> Part 1, lines 163–4

Only in going through clinical depression did I learn of the commonness of mental disorder and realise that I'd lamentably failed to perceive its presence in others, again and again. Here are two instances among many. They were talented people whom I knew a little and spoke to from time to time, and whose sufferings I had absolutely no notion of. They did their best to hide their depression; I wasn't astute enough to see it in front of me. I hope that if I'd known them even a little better I'd have sensed their suffering, but I fear that I can't be confident of that.

'We don't kill ourselves. We are simply defeated by the long, hard struggle to stay alive,' wrote Sally Brampton, the *Sunday Times* agony columnist, in 2008 in her memoir of depression. I corresponded with her a bit on work issues. We once had a disagreement on social media about some impossibly trivial issue. The next day she decided she was wrong and sent me not one but three gracious and totally unnecessary messages of apology. They exemplified a generous nature and also a ruthless self-criticism. I only properly grasped the second of those characteristics when reading in 2016 that she'd drowned after walking into the sea near her home on the south coast. Her *Times* obituary recounted that by the late 1990s she was, on the face of it, enjoying a life of journalistic success and material affluence, with a large house in north London:[1]

> In reality, however, she was grappling with severe depression, which worsened in 2000 when she learnt of the death of her friend Paula Yates. Shortly afterwards, she lost her job as editor of the women's magazine *Red* – she had found it difficult returning to an office after ten quiet years of writing at home, and disliked the management's obsession with celebrities. Around the same time, her marriage broke down. What followed was a haze of alcohol, countless types of antidepressants – all of which were ineffective – and several suicide attempts.

Stephen Tindale was a prominent environmental campaigner and former head of Greenpeace in the UK. His willingness to think freely drove him to support nuclear energy, a stand that cost him many allies in the environmental movement and some friendships. I knew him on and off for thirty years. He came to my flat for dinner. We would meet at weddings and social occasions, and chat about the state of public policy. I had no idea he suffered all the time from depression. It was so oppressive that he attempted suicide in 2007, which left him with brain damage and caused him to walk with a stick. He took his life a decade later. Noting his thoughtfulness and the scorn of his former allies, the *Times* obituary said: 'It is not known to what extent the opprobrium heaped on him affected his mental health, but he continued to suffer from severe depression.'[2]

I can't begin to conceive of the anguish these good, talented and articulate people endured, and how it eventually overwhelmed them. All I can do is at least recognise the state of dislocation and despair that came to dominate their lives. It came to me too, and I had the random good fortune, with the care and concern of others, to find my way out again.

Obstacles to understanding

There are stubborn impediments to recognising depressive disorder. The one that concerns me most is the blizzard of misunderstanding that surrounds public discussion of depression and the consequent difficulty that many sufferers have in owning up to it. Here are one or two examples taken almost at random (in the sense that I happened to notice them, not that I sought them or made any sort of quantitative survey of newspaper coverage).

A minor media controversy erupted in 2019 about comments made by the television presenter Piers Morgan on the ITV breakfast show *Good Morning Britain*. In conversation with a mental-health campaigner, Morgan said: 'I think we need to teach mental strength and we need to teach kids to toughen up a bit . . . life is actually about the real world.'[3]

It's the task of an interviewer to ask challenging questions, but this was marked as an expression of Morgan's own opinion, broadcast to an audience of millions. It conveys the linked ideas that mental disorder is a sign of weakness and that the remedy is in the sufferer's hands. The message is to toughen up and be less sensitive to hardships and disappointments. Morgan means well but has not understood what depression is. And this is now, in the twenty-first century, on national television from a well-known journalist. It's not a fringe opinion in an obscure venue, or an atavistic attitude from a distant and less enlightened era.

No one would tell a sufferer from a disease of an organ other than the brain – someone, say, with emphysema, which attacks the lungs, or lymphoma, which is a cancer that attacks the lymphatic system, or pancreatitis, which inflames the pancreas – to just pull themselves together. The same is true with diseases of the nervous system that are brought on by damage to the brain, such as strokes or Alzheimer's. But depression gets treated differently. This feeds into a common perception among sufferers that they are not really ill, or that there is no prospect of recovery, or that they don't deserve care or treatment, or that they are experiencing an affliction for which they bear responsibility. To justify this divergence of approach, some critics charge that the citizens of affluent western societies are infantilised by a 'therapy culture'. In this modern narrative, they argue, life's normal hardships of sadness and setback are exaggerated and their sufferers are flattered by being depicted as bravely battling against illness. Not only is this fable false, according to the common-sense thesis, it's also debilitating.

Hostility to 'therapy culture'

The account I've just given may seem like a caricature, but it's an exact description of the argument of Frank Furedi, a sociologist who has been given extensive space to expound it in the media. There is, Furedi says, 'a new narrative of illness [which] does not simply frame the way people are expected to feel and experience problems – it is also an invitation to infirmity'.[4]

This notion crops up again and again in discussions of depression, and it is wrong. Furedi's thesis is remote from the reality of patient care. Far from there being a 'new narrative of illness' in the media and public life, depression and other mental disorders are commonly regarded as somehow not real. They are instead, so the argument proceeds, manufactured ailments of affluent societies and privileged people that would be unrecognisable to those suffering real hardship amid famine, oppression and war.

In the coronavirus crisis of 2020, Furedi's disciples believed their time had come. Their continuing message is that we need to collectively develop a bit of backbone and tell children that setbacks are part of the stuff of life which we must all get used to. Joanna Williams, a columnist for *Spiked* magazine, wrote with mock appreciation that the lock-down might accidentally have a bracing effect: 'Ironically, one unintended consequence of shutting schools might be that children have the opportunity to forget directives that everything in life is to be considered stressful.'[5]

Furedi acknowledges that the goal of therapy is to make people better, yet he oddly can't give credit for this. Therapy, in his view, saps an individual's responsibility for their own fate: 'In reality, though, the rhetoric of therapeutic self-determination never granted individuals the right to determine their lives: self-discovery through a professional intermediary is justified by the assumption that individuals are helpless to confront problems on their own.'[6]

This critique of the 'therapy culture' isn't really about medicine or the mind at all. It's part of what's come to be known as the culture wars, though the metaphor is misplaced as only one side is engaged in fighting them. This 'war' is in reality a protest by people who regret social changes that have occurred throughout the advanced industrial economies over the last half-century or so.

The character of western societies has altered since the great legislative reforms of the 1960s and 1970s on divorce, abortion, contraception, race and sex discrimination, and homosexual equality. There's been an expansion of personal liberty and pluralism, and a general decline in religious observance and of the habits of class deference. A determined minority object to the weakening of traditional mores and a supposed unwillingness to hold people accountable for their failings and transgressions. This view tends to be disguised these days in complaints about political correctness, overprotectiveness and the decline of moral and educational standards, rather than being vented in overt prejudice. The notion that mistaking discontent for disease will sap public vitality and responsibility is an extension of this instinct. It fits right in.[7] It's a short step from this sort of premise to expressing scepticism or even mockery about the plight of those who suffer from depression. In a column in the *Daily Mail*, a newspaper reaching deep into the mass market, including many people who will have known mental disorder, Janet Street-Porter has written disparagingly of what she calls 'the misery movement' of women in the public eye who have spoken candidly of their experiences of depression. To her it's all merely a fanciful trend – 'the latest must-have accessory is a big dose of depression' – which sufferers can avoid by adopting her own robust no-nonsense approach to life's pitfalls. She grants that clinical depression is a real condition but, as she gives no evidence of actually believing this proposition, it appears to be merely a rhetorical device to head off accusations of tastelessness before embarking on the serious business of piling obloquy upon famous women such as Emma Thompson who publicly acknowledge having suffered depression.

Street-Porter 'refuse[s] to accept this notion that a whole generation of women are being laid low by an unexplained epidemic of depression', and she gives a rallying cry: 'I find something very slightly repellent about this recent epidemic of middle-class breast-beating. This tidal wave of analysis about why "having it all" isn't what it was cracked up to be. Why daily life is a series of disappointments. Why sufferers feel empty and suicidal. Get a grip, girls!'[8]

Get a grip. *Get a grip*. It's an exhortation directed at those (not only 'girls' but, we can reasonably infer, guys too) who, in the unsparing gaze of the oracles of punditry, fail to show sufficient strength of character and resolution. And it's expressed even by some who have suffered severe mental illness themselves. When the actor Stephen Fry spoke of his own experience of bipolar disorder and suicidal impulses, *The Times* ran an interview with his fellow actor and television presenter Bill Oddie, who said: 'Despite having had my own experience with the condition, I fear that it has become something of a "fashionable" condition to have in this day and age. It is a serious condition, but suddenly people are making careers out of it. Stephen Fry brought a lot of attention to it, and Ruby Wax, but I don't think that the life of a celebrity can be compared to, or relatable to, a normal, everyday life, so I don't know how much good it really does.'[9]

How close this is to the 'get a grip' incantation. There's a ritual assertion that depression is a real condition, which is immediately countermanded by a gibe at those who suffer from it, have a public profile, and choose to talk about it. Depression, you see, may exist in the diagnostic textbooks, and perhaps some people who've borne unimaginable hardships and griefs can be forgiven for succumbing to it, but

– *come on* – it can't be that common, can it? I mean, it's become a fashion accessory; people make careers out of it; they talk about it incessantly; it's part of today's culture of self-absorption; can't they show some decorum and just pipe down a bit?

The difficulty of admitting depression

These critics do make a serious point, in that inflation of an illness can lead to its devaluation in popular understanding, yet I do not share their approach. If western societies take more seriously the afflictions of the mind now than they did in past generations, then this is an unalloyed gain. It helps the sufferers by impressing upon them that they are under no social obligation to hold their silence. It makes society a better and more civilised place. I'm especially thankful when celebrities like Fry go public with their experience of mental illness. So far from being fashionable, it's a difficult decision to acknowledge – to yourself, never mind to others – that you have, according to a traditional way of thinking, shown weakness and (if you're a man) failed to meet the social norms of masculinity. I can imagine that it's especially hard for people who are not only public figures but famous and instantly recognisable.

For the families of those who suffer from bouts of depression, there's also an acute sense of failure mixed with indignation. Alastair Campbell, who served as head of press and campaign strategy for Tony Blair's government, fell into severe depression on leaving Downing Street. His daughter, Grace, later wrote: 'At the beginning, we didn't know how to make him better, which made us feel like total failures. It was impossible for us not to take the fact that we couldn't lift his spirits personally.' Recalling that her father could nonetheless still give every indication of normality in his work and in public, she added: 'I remember thinking how awful it was, that he couldn't bring up his spirits for us but he could for them, the politicians. I've learnt now, in retrospect, this was because he found work an easy place to escape to when he was depressed, whereas he found being vulnerable in front of us painfully hard.'[10]

It's a mountainous task to carry on as if nothing has happened and all is as it's always been. If you can manage to give such an impression, it's an act. It's a performance for those who see you in the strictly delimited part of your life, namely your work and public persona. I did the same during depression. It's not possible to maintain a similar appearance of normality in private and among those who know you best. To be vulnerable and needy before them, and especially those who in normal times depend on you, is unavoidable. It thereby compounds the sense of failure and guilt that you carry everywhere. Friends and family in turn feel a sense of betrayal (the word is not too strong) that your more lucid and apparently stable moments are reserved for others, including strangers. And so the vertiginous descent into depression and despair takes a further twist.

Obstacles to parity of esteem

It ought not to be like this. In recent years, public-health policy has incorporated the principle of 'parity of esteem'. This means that mental health must be accorded equal priority with physical health. It's not just a stated objective but a legal requirement of health provision, included in the Health and Social Care Act 2012. If we are to get anywhere close to this in practice rather than rhetoric alone, and to alleviate the bane of mental illness, then we need to talk a lot more about the subject rather than less. Modern Britain does not have a 'therapy culture'. Nor does any other advanced industrial economy I'm familiar with. But if such a thing did exist then it would still be more valuable and humane than a bleakly indifferent culture. Above all, we have to dispel the notion that mental disorder is somehow less real than a physical disease or injury. And that will be hard, as there are stubbornly held assumptions in public discussion about what illness really is.

Here's a case in point. It's from the early years of this century, but I know of no evidence that the attitudes it records have substantially altered since. In a study by Jon Stone, Professor of Neurology at Edinburgh University, people attending an outpatient department were presented with a hypothetical case of someone going to the doctor with a complaint of weakness in their legs. The interviewers asked the respondents what they thought of various alternative diagnoses. Those diagnoses that suggested the mind was responsible (for example, in producing a psychosomatic effect or in weakness due to depression) were regarded far less sympathetically than explanations to do with physical malfunctions like multiple sclerosis, functional weakness or a stroke. With a diagnosis of a mental condition, the complaint was commonly regarded as imaginary and not a reason to be off work – and indeed not a medical condition at all.[11]

There's a cliché about mental illness: it's all in your head. The implication of this purported witticism is that, however vivid mental problems may appear to the sufferer, they're not really there. There's no genuine cause for them. The remedy is hence in the hands of the sufferer. Pull yourself together. Face up to life. Get a grip. *Get a grip*.

I never reasoned it quite like that but, till I experienced it myself, mental disorder was still for me an opaque concept. I thought of it with, at best, tacit bemusement. My thinking was crude and it ran something like this. How can an ailment for which there is no external cause be real? If there's nothing in your life to be sad about or to make you fearful, how can you genuinely be depressed? And if there is some identifiable cause for melancholy, then the sufferer has to accept that this is the lot of humanity. It always will be. As the essayist and philosopher Miguel de Unamuno wrote: 'There is something which, for lack of a better name, we will call the tragic sense of life, which carries with it a whole conception of life itself and of the universe, a whole philosophy more or less formulated, more or less conscious.'[12]

Like Janet Street-Porter, I could grant the theoretical possibility of clinical depression but would implicitly acknowledge it as worthy of sympathy only for those who'd endured danger or tragedy, perhaps by being shot at in war zones or suffering the wrenching loss of the death of a child. There were, on my unstated criteria, deserving and undeserving sufferers from depression according to the privations they'd experienced. And if there was no proximate trigger for it, then the depression could surely be dispelled by the merest assertion of common sense. The disorder might be genuinely distressing, but it was at root illusory. At some point we must all adjust to the tragic sense of life; it's part of our growing up, as individuals, a society and a species.

If mental disorder were directly observable, like a physical wound, no one would doubt its severity. Yet many people are dismissive, and some of them are medically trained. When I was in my deepest pit of depression and desperate to find out more, I read a *Times* review of a book I've already cited by the psychologist David Clark and the economist Richard Layard. These authors urge much greater use of psychological therapies in public health care. The reviewer, the conservative social critic Theodore Dalrymple, was unimpressed with their case. He scorned 'the authors' extreme naivety, which is refreshing almost in its innocence . . . For the authors, such phenomena as anxiety and depression are facts of the same order as earthquakes and hurricanes, and they ignore entirely the evidence that they have largely been constructed by, among others, the purveyors of useless but expensive drugs.'[13]

Dalrymple is a trained psychiatrist. He must know there is no shortage of evidence that depression and anxiety are real, and that antidepressant drugs are effective (they are also, by the way, inexpensive). The notion that Big Pharma has constructed a mythical illness in order to sell placebos at a high mark-up would be risible were it not for the destructive consequences that this conspiracy theory has for public understanding of devastating illnesses. Dalrymple's invective belongs in the same category of myth as the ones pushed by the anti-vaccine movement, whose calumnious efforts to malign the developers and manufacturers of the MMR vaccine have reintroduced measles to countries where it had been thought eliminated.[14]

The fallacy of dualism

The fallacy that there is a separate entity of mind, as opposed to the real stuff of medicine, is pervasive in our society. I at least tacitly held to it myself once. The only way in which I can prettify my misunderstandings is to offer an evasively creative reinterpretation. To say that depression is all in the head is literally true, so long as we clearly understand what that means. There is no real division between mind and matter, and mental disorder is inseparable from the workings of the brain.

The reason why mental and physical health merit parity of esteem is not mere politeness – the notion that sufferers from mental disorder should be accorded the diplomatic fiction that they have a genuine illness, in the hope that they will thereby respond more quickly and positively to the placebos of pills and talking therapies. Parity of esteem in the treatment of mental and physical illness is not an exhortation but a description of what it takes to provide effective care.

That's because the body is the sum total of our existence. It's all there is to us humans. I'm not saying merely that the life of the mind depends on our bodily existence and would be dead without it. That's true but not profound, for it skirts the fact that everything we are, including our thoughts, emotions and memories, is a product of material processes. Certainly, medicine is more than applied physiology. Its practitioners treat disorders of the person, which can involve surgery but also such apparently mundane tasks as sorting out a patient's welfare benefits, and everything in between. It is a person who suffers clinical depression, not a brain. Some psychologists go further and argue that neuroscience is not necessary in the task of understanding cognitions, affect and behaviour. Yet I'd contend that mental health is worth the same as physical health because, at root, it's the same sort of thing.

The notion that depression is a voguish medicalisation of the messiness of human emotion and the normality of sadness is widespread. I've read these assertions of bluff straight talk, whose authors imagine they're struggling against the obfuscations of clinical jargon and the tendency to see illness where there is only sadness. Their argument from common sense fails every time, above all because it is just not very sensible. It isn't founded on facts. It deals with the countervailing diagnostic and clinical evidence and numerous studies by ignoring them. Above all, it is prey to the fallacy of dualism.

The objection to 'wide' definitions of depression

There is still, even so, a scientifically responsible case that mental disorder is a shifting category. It doesn't depend on the fallacy of dualism. Its advocates maintain, rather, that since the 1980s the diagnosis of depression has expanded beyond the limits of usefulness, by including not only severe depressive disorders but also more normal states of intense sadness.

This argument is made by two American sociologists, Allan Horwitz and Jerome Wakefield, in their book *The Loss of Sadness*.[15] The book's thesis is that, while depressive disorder certainly exists, there is not a global pandemic of it. Rather, the numbers of sufferers have been artificially boosted by including within the label of depressive illness people who experience episodes of more conventional sadness. The authors maintain:

> Before 1980, for the 2,500 years since the dawn of psychiatric medicine, only symptoms that were 'excessive' and inexplicable relative to their provoking context were considered to be signs of a depressive disorder. After 1980 all symptoms, even those that are proportionate to their provoking cause, were defined as disordered. This change means that intense natural reactions to loss events as well as disordered responses have been seen as mental disorders, thus accounting for the apparent increase in depression in recent years.[16]

Proportionality is their point. A mental disorder is a response that isn't proportional to any presumed cause. Depression is a condition that goes far beyond the external trigger for it, if indeed there is one at all. A definition of depression that obscures the condition's abnormality isn't useful. The critique that depression can be too broadly defined needs to be taken seriously. The work of Horwitz and Wakefield, and others, contributes to a genuine debate about modern diagnosis and treatment.

So do the writings of Steven Pinker, who in his book *Enlightenment Now* argues that there is a 'spiral of recursive improvement' in human affairs. Against the notion of an epidemic of mental illness in modern western societies, he notes that the list of mental disorders in the *DSM* of the American Psychiatric Association tripled between 1952 and 1994 to include almost 300 such conditions, among them avoidant personality disorder (which might once have been termed shyness). Pinker comments: 'By the same shift, the label "depression" today may be applied to conditions that in the past were called grief, sorrow, or sadness.'[17]

Some radical critics believe that the individualistic ethos of market economies spreads mental disorder and makes people ill. However, the evidence they cite is generally purely anecdotal and subjective.[18] I've even come across the charge that depression is a tool wielded by the forces of political reaction in order to suppress dissent and enforce control of a compliant population. The essential truth in the objections to what I'll call the 'wide' diagnosis of depression is that, while modern western societies have many flaws and pathologies, they are not a dystopia. They are constitutional societies in which, broadly speaking, imperfect and sometimes remote public bureaucracies do a substantial amount of good. When 'depression' becomes a synonym for disagreeing with the policies of the government of the day, then the shift that Pinker refers to really ought to be corrected in public discussion.[19]

It isn't part of my argument, or necessary to it, that depression is becoming more common. It's enough that depression as a medical condition is already widespread, real and devastating, and that it is not sufficiently understood. A rise in reported cases of depression in affluent societies may perhaps cover a range of mental conditions, but the recognition of this disorder is a humane advance in the history of civilisation and an expansion of knowledge.

Here's an instance. The *DSM* has expanded its criteria for diagnosing depression to take account of whether or not the condition is chronic. This is a change in classification as it goes beyond distinguishing only between mild and moderate to severe cases of depression. And this is an advance in diagnostic precision (as well as compassion) because it recognises that mild depression can be a persistent depressive disorder in the same way as severe depression is. We now know more about depression, including the fact that the condition is not necessarily self-remitting but can persist for long periods and even for life.

Even if it were true that depression is being diagnosed too readily in western societies, this would be a benign error. The social consequences of such a mismeasurement would be far less damaging than if depression were being under-assessed, let alone disputed and derided. Awareness of mental disorder is an example of greater clarity and concern in the twenty-first century about the extent and nature of human suffering. As Pinker puts it: 'The expanding empire of psychopathology is a first-world problem, and in many ways is a sign of moral progress. Recognizing a person's suffering, even with a diagnostic label, is a form of compassion, particularly when the suffering can be alleviated.'[20]

We need to talk about depression

The diagnostic labels have fluctuated over generations. Whereas W. H. Auden wrote in 1947 a long poem titled *The Age of Anxiety*, which inspired a symphony by Leonard Bernstein, we now customarily speak of depression rather than anxiety. The labels tell us not that we are mistaking normal human problems for illness, but that language shifts. The phenomena these terms identify, of depression and its consequences, remain the same. Lives, livelihoods, families, wealth and welfare are destroyed by depression. It's a ruinous affliction. There's a pressing need for greater public education and understanding about depression, and for the stigma attaching to the illness to be dispelled.

Far from there being in modern societies a culture of resorting to therapy, depression and other disorders are not talked about anything like enough. Sufferers are commonly silent about their plight and resigned forever to their fate. I was of their number but had the opportunity, as a journalist and author, to be voluble instead. Resignation is not necessary. Depression can be treated, once it's recognised for what it is.

PART TWO:

TREATING DEPRESSION

5

DIAGNOSING DEPRESSION

> His misery . . . swells above all the hilles, and reaches to the remotest parts of this earth, Man; who of himselfe is but dust, and coagulated and kneaded into earth, by teares; his matter is earth, his forme, misery.
>
> John Donne, Meditation VIII, *Devotions upon Emergent Occasions* (1624, first edition)

Every sufferer's experience is different, but the symptoms are recognisable. Mine were obvious to anyone trained in medicine. But it took a while for me to get that diagnosis and I didn't understand what was going on. This is what it was like.

For a start, I have to define it negatively, in the sense of what it didn't resemble. It wasn't at all like the sadnesses, setbacks and disappointments that conventionally punctuate life and of which I'd had many. It was neither a nagging sensation at the back of the mind nor a seasonally recurring disorder. It came with no warning and it superseded every other thought and sensation. Nothing else can occlude severe depression; it allows no moment of levity or recreation. Like vertigo it supplants every other sensation with fear, yet with depression the sense of falling isn't even broken by the ground below. It's a descent into unfathomable darkness where you can have no conception of a physical or temporal end. Though its symptoms may include suicidal thoughts, depression is more severe even than a conviction that life is no longer worth living. Instead, it casts its shadow over the whole of life, retrospective, present and future. It impresses upon you that every advance to date has been futile, and this realisation, this self-knowledge that has never previously been grasped, is on its own a debilitating burden.

Indeed, to the sufferer of severe depressive disorder, life has been worse than futile: it's been misunderstood, as in a dream, except that others have not been in the same state of illusion about you. They know instead your every failure and they observe, perhaps with sadness but in any event with a sense of inevitability, your decline and descent. It is the state of Gilbert Pinfold in Waugh's novel, on fearing his madness and hearing voices that he can't fathom: 'And in that moment of agony there broke not far from him in the darkness peal upon rising peal of mocking laughter . . . It was not an emollient sound. It was devoid of mirth, an obscene cacophony of pure hatred.'[1]

For some, suicidal thoughts arise through the public revelation of a shameful secret. For me, depression created such thoughts rather than being elicited by them. It was the trigger for a personal epiphany. Here was a moment of instant clarity, after a lifetime of ignorance, about a shame that couldn't be extinguished. It was a terrible debasement, in which I saw myself fully and accurately for the first time. I'd been blind to the presence of those who depended on me. For a tragic figure, for the shamed Othello, the question was: 'But why should honour outlive honesty? Let it go all.' For me, a mediocrity in everything except the extent of my obtuseness in recognising where it had all gone wrong, the shame was not measurable in the same way. I didn't have even the defence of a prior belief in having been wronged by those closest to me. Nor was there any hope of moral recovery in the hereafter. There was no hereafter; there never would be.

The human yearning for fellowship

Religious beliefs have no hold on me. Nor do I regret my inability to affirm creeds about the origins, end and purposes of the universe that have no evidence to support them. Nor can I even take them as elevating metaphors to live by, for they are not. In the Book of Judges (11:31), Jephthah vows to sacrifice 'whatsoever cometh forth of the doors of my house' on his return if God will grant him victory over the Ammonites. He at least has the decency to rend his clothes when this turns out to be his daughter. In Byron's telling, she urges her father on: 'Since the triumph was bought by thy vow, / Strike the bosom that's bared for thee now!'[2]

There is no beauty in this imperishable image of fanaticism. I have no regrets, even momentarily, for my conviction that there's no truth in it either. Yet there is a dogged justification for the resilience of religion, even while its doctrines become ever less credible to the critical mind. In navigating life, we find it hard to do it alone. Religions are unalike in symbolism and dogma, but they offer similar comfort to the faithful in acclimatising them to the journey of life. People seek religious fellowship out of universal and not only personal needs. It's natural to wish to feel part of a wider community, gain a sense of purpose in life, crave forgiveness for the things we've done wrong, find answers to cosmic mysteries and – for some – have the assurance of conquering death. These are not ignoble motivations and all but the last would remain powerful human urges in the absence of religion.

The hope of personal immortality may be a widespread wish but it is not a universal need, and is an appalling prospect once you try to reason what it would be like. No religious thinker, ever, has managed to describe how eternal life could be joyful. Fellowship in this life, on the other hand, is a solace that few can manage without. As the novelist John Cheever, who clung to Christianity, put it in the words of the protagonist of his short story 'The Housebreaker of Shady Hill': 'It is not, as somebody once wrote, the smell of corn bread that calls us back from death; it is the lights and signs of love and friendship.'[3]

It was what I craved but couldn't have. I tried to imagine the burden it must be to know me while discreetly having to pretend not to know of my failings and transgressions. Much later, when I determined to find out all I could about depressive disorders, I learnt that the corrosions of shame are frequent and that they make recovery hard. Receiving well-meant invitations to meet and socialise, depressives may fear that these are extended out of politeness and pity, and that the shame that accompanies them is universally known. The kindest course, when offered the solace of companionship, is often hence not to accept and not to explain.

I was desperately fortunate. Close friends came together to ensure that I would not be left on my own. They planned, and allocated among themselves, the thankless daily task of meeting me at lunchtime or calling on me in the evening with food they'd bought or cooked for me. They'd have me to stay for part of every week to make sure I ate and slept and was capable of travelling to work.

They arranged this in the first instance spontaneously. I reached home early one evening and sat on the steps to my front door to greet the cat from the downstairs flat. After a few minutes, it had become appreciably darker and it seemed that only agonies awaited me across the threshold. These weren't nameless, formless terrors of the type that children imagine after the lights go off, but the agonies of a mind locked in rumination and reproach, and I couldn't face what lay beyond. What lay ahead was a tormented evening and an endless sleepless night. So instead of opening the front door, I walked unsteadily to the bus stop, sprawled on the back seat and came out at the terminus at Finsbury Park in north London. I waited there for an hour or so, till my friends were home a little way up the hill. They offered food but I couldn't eat; I crumpled into the chair, and felt it would be an insurmountable effort to speak. They sat with me.

It became a routine. For the first part of the week I'd stay with them, a married couple whose daughters, one at university and one doing A-levels, I'd known all their lives. As our respective offices were all within walking distance, I'd travel on the same train with these long-suffering friends in the mornings, at what would be an unusually early hour for the common stereotype of a journalist. It didn't matter; in normal times, I get up at around 5 a.m. and write, which is my most fluent time for working, and subsist on coffee. Much later in the morning, verging on lunchtime, I have breakfast. In those days, my routine altered. I wouldn't sleep and my friends would ration the coffee in case (as was likely) it was aggravating the problem.

Fears of the disordered mind

Theirs was a noble and selfless attempt to provide me with a sanctuary in which to recover my sanity. Yet for all their care, my friends couldn't displace my stubborn, ineradicable certainty that I was monstrous. One evening after dinner, in their sitting room, I asked if they thought I was evil. They politely laughed, as if I'd made a joke, and then stopped in bewilderment. It was a question I had to ask as I badly wanted to know the answer. But none they could give me was of any use, as I already knew the truth and was filled with horror at it. That realisation was most acute in the twilight hours, when I was awake but fitfully slumbering and then wide-eyed again. The cycle began again every few minutes.

For no reason I could adduce, the image of a windowless round tower and an earthen floor kept coming to me, night after night, in these moments, presumably first dreamt in the rapid-eye-movement stage of sleep, and then consciously recollected just after the same state on other nights. Few psychologists take seriously the Freudian notion that dreams are unfulfilled unconscious wishes, but there is much debate about why they happen. The Swiss psychoanalyst Carl Jung theorised, like Freud, that dreams are messages from our unconscious, but he concluded that they reveal essential information about the cure to our internal conflicts. One influential theory, advanced by the neuroscientist Allan Hobson, is that dreams are in fact random images stitched together by the brain. That explanation makes intuitive sense to me and it at least damped down my anxiety at the time. I'm drawn to the idea that grand theories of the Freudian and Jungian type are undermined by a more modest explanation that focuses on the complexity of the human brain.

A random image was what it was, but I overlaid it by imagining the structure toppling in on me while the world beyond was invisible and unperturbed. This sort of grasping for significance is what we humans are like. We constantly seek to understand. That urge has a beneficial expression, in the search for knowledge, but also a less benign one of uncritically accepting explanations that satisfy our wish that the world should be a particular way. A small example of the latter tendency is the thinking of the conspiracy theorist. You'd imagine that someone who believes the social world is controlled by a malevolent conspiracy of the deep state would live in constant fear for their life, as the conspirators would presumably not allow the exposure of their misdeeds. Yet my experience of conspiracy theorists, on, say, the link between the coronavirus and 5G mobile technology, is that they can't stop talking about the subject. The belief, indeed the certainty, that they possess esoteric knowledge appears to be the real attraction here.[4]

Human relations do not conform to such simple patterns. Our lives are a collection of essentially chance occurrences on which we seek to impose order and narrative. We look for a story, whereas the narrative we create is only what we derive from looking back, not what we experience at the time. Our search for understanding accords with a metaphor coined by Otto Neurath, the Austrian philosopher, about scientific knowledge: 'We are like sailors who have to rebuild their ship on the open sea, without ever being able to dismantle it in dry-dock and reconstruct it from its best components.'[5]

The tumbling tower wasn't a sign or prediction; it was more a wish for an end and an indication of a weight of shame. I carried this burden with me wherever I went. One day I noticed in the office a leaflet pinned to a noticeboard. I forget the message, but it gave the phone number of a counselling hotline. It had probably been there for some time as a guide to employees on how to manage worries about money, marriage and mental difficulties, and I'd not noticed.

I took down the number surreptitiously, for it seemed a moment of weakness. This was not the quality of the Stoics. But when I was back at home (for I was determined that I would only do so at my home, not at my friends' house, and only when there was no chance of being overheard) I phoned the number. My usual manner, on a public platform or on the radio, is to talk. And here the words wouldn't come, just choking sobs, on and on.

The friendly voice at the other end waited with inordinate patience. I stammered eventually that I was having a difficult time and was suffering from stress. I didn't know what stress meant as a condition, but it seemed the least-bad summary I could come up with for why I was in such a state. When a modicum of composure returned, the voice took me through her checklist of symptoms. She asked in particular if I was at risk of self-harm. It seemed a strange question as the very possibility hadn't occurred to me, but I found I wanted to talk about the darkness and the mental images. I had no inkling what I could say about it all, though. It would sound insane.

She did her best, her very best, in attempting to calm a disturbed person and assessing whether I was suicidal. But in retrospect I suspect there was only one remedy that she had to hand and that it would be the same for everyone. She emailed me the next day with a referral to a therapist. That was a new one to me. I had no inclination to find myself through psychotherapy, and no belief that fishing through my memories would tell me anything reliable, let alone useful or interesting. But I was desperate to talk to someone; and my friends, knowing that I was not myself and unable to get me back again, thought it sensible that I meet a dispassionate professional outsider. So I phoned the therapist and made an appointment. She detected a note of urgency in my voice and advised me to visit a GP. That was not only good advice but, sadly, the only worthwhile thing I got out of the many hours in which we talked.

Therapeutic explorations

I turned up one afternoon for my first appointment. The therapist lived and worked in a red-brick Edwardian block in north London. She was a slight woman in late middle age, in carpet slippers and with a shock of grey hair. I wanted to like her, and her manner was reassuring. Seeing I was in distress, she suggested we meet three times a week over the next couple of months and then take stock. I readily agreed, believing I'd found a confidante, and told her of the things that had brought me low.

They seemed mundane when spoken out loud, and my concerns bizarrely exaggerated. Again it struck me that the ebbs and eddies of existence are what everyone experiences, and that to many the image of maturity in Thomas Cole's *The Voyage of Life* was common too. We are, at that stage of life, responsible for ourselves and often others, and to fail in it is dismaying but not exceptional. In talking of my short-comings, I could recount the loneliness, guilt and ennui but not the darkness they brought in their wake. It was all so dramatically overstated. My shame erupted further at my inability to keep in proportion the disappointments, while also compounding the sense that my failings were far worse than I'd yet realised. On departing my first session with the therapist, I extended my hand – which she was extremely reluctant to take. It surprised me, but I realised in that instant that she believed she was applying a professional method of greater significance than the mere acts of listening, talking and advising.

So it went. I would turn up and the therapist would wait for me to begin. That too, I belatedly realised, was part of the technique. It wasn't a conversation but a staged exercise. The aim was not to guide or provide a template of advice but to explore my thoughts and delve into them. There was a principle to it as well as a method. The therapist explained that the reasons behind my state of distress would be uncovered by exploring my feelings about past events. As I talked and she listened, we could identify persistent themes in my life that had caused my present mental disorder.

The therapist's intentions were good-hearted, but it was a destructive encounter. Not going at all would have been a better idea. If you start from the belief that, in therapy, there is some catalyst waiting to be uncovered you'll look for it in every word or recollection. I knew she was looking for it, and she knew I knew she was looking for it, and so it went. Sometimes, to throw her off the scent, I wouldn't start talking but would just wait for her to begin. Eventually she would, with a few words asking how I'd been the previous day or two. And then we were in the same position, where she waited for me to recount my mental states.

Sometimes the exchange was bleakly comic. For one morning session I'd arrived early and hence went to a coffee shop over the road. I don't have a sweet tooth but on this occasion, for some reason, I felt like having a pastry. So I bought one and sat down. I ate it all. It wasn't bad but it was sweet, which indeed was the point of it. I made the mistake of mentioning this to the therapist when I got to her twenty minutes later, and we did an elaborate verbal gavotte around the presumed significance of this decision. Was I, she speculated, at last allowing myself to indulge in sensual pleasures as a reaction to my self-castigation? Was the pastry, perhaps, emblematic of some yearned-for state that I'd denied being deserving of?

The answers to these questions were, I knew, no and no. The pastry was not a symbol: it was a pastry. It had no wider significance. But I couldn't convince her. Nothing could convince her that my decision to buy and eat a pastry was an essentially meaningless occurrence driven by my brain states. Eventually we argued it out and she left the matter there. But it stayed with me as an indication of what was going on in these sessions. For me, it was a chance to talk to someone who I hoped would have wisdom and insight and, above all, a cure for what I was feeling. For her, it was more like a fishing expedition. She would trawl through my anxieties and recollections, and indeed anything I had to say, seeking to apply it to my current frame of mind. She would alight on phrases and reminiscences that she hoped might yield information if she only pushed me a bit harder to explain them.

Far more seriously, in recounting my life's experiences (not something I wished to do, but it was information she sought) I recalled the suicide of a schoolfriend when he and I were sixteen. It was at the time traumatic and deeply tragic, and it had nothing to do with the mental turmoil I was now experiencing as an adult in middle age. I knew this. Thinkers have wondered for millennia at the continuities of personal identity through a lifespan that define the human condition, but these questions were of no relevance to me, then and there. I was, for all practical purposes, not the same person that I had been decades before. I didn't wish to dredge through distant memories and I resented the waste of time expended in doing so. All I wanted was to get out of the pit, but instead its walls stood sheer above me.

The therapist assumed that my scepticism was a symptom of the problem, as I was showing 'resistance' to opening myself up. It was infuriating. I couldn't see a purpose to the discussion or any prospect of a cure by this route and asked that we terminate our sessions. Even then she wanted one more, impressing upon me that we needed this additional session in order to recap all that we'd discussed. We didn't but I limply assented, returned a few days later, predictably got nothing out of the encounter except an implicit scolding for my lack of faith in her methods, and then thankfully escaped for good. Her parting words were that she was sorry not to be seeing me again because there was a nice person inside. The caveat would normally have made me laugh. It was a model putdown.

I never saw her again. I just paid the bill.

The limitations of the generous impulse

While the therapist was trying to work out what was going on in my mind, I was wondering the same about hers. She was trying to apply, I later learnt, a technique known as psychodynamic therapy. The 'dynamic' part of this noun phrase indicates a focus on the various factors (or 'dynamics') that affect a person's life and that may be responsible for their current difficulties. The idea is that the therapy will examine these factors, some of which may date back to the patient's early childhood, and seek to address how they're causing current mental disorder.

It didn't work for me. My goodness, it didn't work at all. It made things worse. The best I can say is that it was dispensed by a kindly if bumbling practitioner, and that psychodynamic psychotherapy is at least preferable to earlier remedies that are set out in the next chapter.

This was my first experience of psychodynamic psychotherapy, and it will be my last. It wasn't merely the wrong choice. My sessions with a therapist were an education in how a generous impulse to help people with their problems can easily end up doing damage to those with mental illness. I witnessed that potential and experienced that effect. Though I believe that the therapist's handling of my case was disastrous, it never occurred to me to pursue any sort of complaint against her. Why would I and what would be the point? I doubt this therapist was temperamentally suited to her work in the first place, but her sincere wish was to help and the problem was more fundamental than an issue of personality or of the clash between hers and mine.

The beginnings of discovery and diagnosis

I had nowhere left to go. It seemed to me that I'd conscientiously sought help yet been beyond it when it arrived. Having run aground on the shoals of life, I'd been unable to clamber back even to sanity, let alone equilibrium and contentment. The conviction that my mental disorder was born of unparalleled human failings was reinforced by my encounter with psychodynamic therapy.

My worries multiplied and burgeoned. One of the many fears I couldn't shake off was that my friends, who continued to give me hospitality for part of every week and then to faithfully visit me when I was at home, were unable to cope with me. After all, they'd encouraged me to try talking with a therapist, and it hadn't worked. They'd hoped for a meeting of minds, where I'd find peace; instead, I'd been repelled by the experience of therapy and fled. This type of apprehension too, I learnt, is characteristic of a depressive state. Why, I reflected, would anyone wish to see me and take me out for a meal? It made no sense. I was a moral outcast. So I ceased replying to messages. If there was a knock at the door, I would crouch low and silently wait till the visitor had given up and departed.

Shut up in my eyrie at home, I wrote to a former colleague I'd worked closely with and who I thought might be able to help break the impasse. Mark Henderson had been science editor at *The Times* before taking up a post at the Wellcome Trust, the medical charity that is a large benefactor of health research. I told him I was unable to comprehend what was happening to me and earnestly asked for guidance on the science of mental disorder. He replied immediately and sympathetically, and put me in touch with two people of great experience in the fields of depression and neuroscience. The first was the redoubtable mental-health campaigner Lord (Dennis) Stevenson, who invited me to see him at his home in Westminster the next Sunday afternoon.

It was an ambitious plan just to get there. I wasn't particularly safe to cross roads as I didn't always notice traffic. For weeks, apart from trips to the therapist, I'd travelled on just two bus routes: between my home in east London to the *Times* offices, and between home and my friends in north London, who I knew would – extraordinarily – welcome me and give me their spare room at any time and in whatever state they found me. But I tried it and eventually made it to the Stevenson family home.

Again, I could scarcely get the words out as I curled up and cowered at one end of the sofa. Lord Stevenson patiently waited for me to tell him my story; then he asked his questions. He explained for a couple of hours the state of the science, the struggles and as yet rudimentary progress of neuroscience in decoding the mysteries of the brain, and the plight of the hundreds of people he'd talked to who were in my position. He asked me what I most enjoyed doing, comparable to his enthusiasm for watching Arsenal play football or listening to Beethoven. I mentioned literature and music and that I wasn't able to do these things any more. 'Well,' he said, 'you almost certainly have clinical depression!' That was his way of distinguishing between mental disorder and conventional sadness. And it was true. I couldn't maintain attention long enough to read a full sentence; and the Beethoven whom I too revered had become a cacophony.

Lord Stevenson wrote to me the next day and phoned me every few days thereafter to check how I was. He still does, years later. I didn't take in much that he said. I just knew there was someone who was interested in where I'd ended up and who wasn't going to judge me. He told me I was brave to admit to depression.

I did admit to it, not through courage but through weight of evidence. I couldn't live normally, and my published output was pitiful. There wasn't much of it and my mind would struggle to recall anything of relevance as I considered the subject matter of public policy or whatever else I was called on to write about. I decided I had to tell close colleagues and senior management before I was exposed as a bluffer and a laggard.

My colleagues, leader writers and columnists at the newspaper, turned out to have already noticed something was badly awry in my work, gait and appearance. They'd covered for me without my realising or noticing it. And my employers were immensely sympathetic. They impressed upon me that I could and should get help. They hence sent me in the first instance to the in-house clinic to see the doctor. His name was Jeremy Nathan and he ran a minor injuries clinic at the office for some mornings each week. I got to know him well over the many months ahead. He was friendly, knew my writings for the newspaper and, like the psychotherapist but to a different purpose, endeavoured to put me at my ease. I explained haltingly and over what seemed like half an hour but was in fact just a few minutes, all that had gone wrong. Again, as with the therapist, I found myself unable to explain the darkness that enveloped me every moment.

Dr Nathan immediately perceived that it was severe depression and, needlessly apologising for simple visual aids, showed me a health information sheet listing the symptoms of clinical depression. It was a revelation. I had the lot: I could have been a chapter, or at least a footnote, all on my own in a textbook of psychiatric disorders. I had sleeplessness, listlessness, hopelessness, sluggish speech and movement, a total loss of appetite for recreation or pleasure, and some things I couldn't explain at all. My problem of short-term memory loss was acute, and the fear of it has never quite left me. (To this day I do almost all my shopping online, partly because of convenience but mainly so that I'm not at risk of inadvertently leaving without paying.)

Dr Nathan was wary when I recounted my experience of psychodynamic psychotherapy and was righteously angry at the therapist's parting shot – which I'm sure was meant with kindness and encouragement but was, on reflection, a measure of our mutual incomprehension. He advised me to take two routes to get better. One was psychological therapy of a different kind. The other, if I wished it and which he would recommend and prescribe, was to take antidepressants. I knew little of either course at this stage; but I badly needed help and agreed.

It didn't feel like a huge advance, but it was the first step to getting well. I had an accurate diagnosis of a clinical condition that meant something, and a plan for how to treat it. To know that it was not just me was a support in itself. Because I earnestly wished to be better, my yearnings focused on the most immediate remedy, the pills, and I built upon them hopes that they would instantaneously restore my faculties.

Of course, they didn't work like that. They are a tool, and for me they were a secondary one, in emerging from depression. That's not to dispute their value. They are an integral part of modern treatment for mental health, which has made great advances in the past half-century but which also contains much that doesn't help. Better treatment of mental disorder is a long-standing quest, and the impulse has been humane even where historic methods seem atavistic or even shocking. In the following two chapters I summarise the historical search for medical and psychological treatments of depression, before returning to my own experience of what helped me cope with the consequences and eventually cured the malady.

6

MEDICAL TREATMENT

> Their counsel turns to passion, which before
> Would give preceptial medicine to rage,
> Fetter strong madness in a silken thread,
> Charm ache with air and agony with words.
>
> William Shakespeare, *Much Ado About Nothing*, Act 5, Scene 1

The mutability of mental illness is far from a new insight. It was because of increasing confidence that the maladies of the mind were capable of being cured that the most disturbing images of treatment stick in our historical memory.

With the dawning of the Age of Reason came custodial institutions (originally established by Louis XIV of France in the mid seventeenth century) for treating insanity. There is a famous image of Bedlam (Bethlem Hospital, rebuilt in 1676) in William Hogarth's depiction of *A Rake's Progress*, painted in the 1730s, which is on display at the Soane Museum in London. The final scene, 'Rake in Bedlam', is assumed by critics to be modelled on two statues by Caius Gabriel Cibber, *Raving* and *Melancholy Madness*, which at that time were part of the hospital gates and are now in the Victoria and Albert Museum. To us these are grotesque and disturbing images, but the madhouse was already a staple of popular and even high culture. Nathaniel Lee, the seventeenth-century dramatist, depicted the inmates in a way that we would now associate with the reputation of Bedlam:[1]

> Like a poor lunatic that makes his moan,
> And for a while beguiles his lookers on,
> His eyes their wildness lose.
> He vows his keepers his wronged sense abuse;
> But if you hit the cause that hurts his brain,
> Then his teeth gnash, he foams, he shakes his chain,
> His eyeballs roll, and he is mad.

It looks, sounds and was a brutal regime. But it has to be compared with what preceded it, which generally involved keeping mentally disturbed people at home, imprisoned and mistreated. There was nothing and no one to stop them being beaten and starved. Even so, institutional life made extensive use of manacles. The French Revolution was genuinely a movement of the Enlightenment in its treatment of mental inmates, with the radical physician Philippe Pinel becoming famous for unlocking chains and adopting psychological therapies. Before him, according to one authority: 'In general, conditions were atrocious. There was no provision for cleanliness or comfort, much less for anything resembling therapy (probably because at the time the insane were considered both insensible and incurable), and asylums were custodial institutions rather than therapeutic hospitals. There were exceptions to this situation, but barbarity and ignorance were the rule.'[2] Pinel was a pioneer in humane treatment not only in removing the barbarities of institutional life but of seeking to understand mental illness by detailed observation.

In Britain, too, the humanitarian impulse was evident both in treating the mentally ill and also in increasingly seeing them as patients rather than as inmates. Quakers in York towards the end of the eighteenth century began campaigning to phase out the chains after the death of one of their number, Hannah Mills, as a consequence of her maltreatment in an asylum in 1790. She had been diagnosed as having melancholia, which we would now call depression, and died only a few weeks after entering the institution. Her friends and relations had been denied the opportunity to visit her. The Quakers established their own institution, the York Retreat, to take care of Friends who suffered mental illness. Those who were not members of this religious denomination sadly still had to make do with York Asylum.[3]

These were precursors for a great expansion of mental institutions in the nineteenth century. The population of asylums was dominated by poorer people, though. For wealthier classes, treatment would be outside the asylum. We don't know how many of those who were committed were, like Hannah Mills, suffering from what we'd now recognise as psychiatric disorders, but there was less stigma for those with money to be treated for 'melancholia' or 'nerves' than for insanity. The family doctor and the revival of the spa were the recourses of wealthy patients, who in time would tend to get better. The mental institution by contrast became a place of incarceration rather than cure, divorced from the Enlightenment ideals that had originally inspired it. There is a direct line between the institutionalised inmates of the mid to late nineteenth century and the fate of Randle McMurphy, the rebellious psychiatric patient who is lobotomised at the end of Ken Kesey's celebrated novel *One Flew Over the Cuckoo's Nest* (and was played by Jack Nicholson in the still more famous film adaptation of 1975).

The tragedies of lobotomy

A lobotomy is a neurosurgical procedure whereby the groups of nerve fibres in the brain's prefrontal lobe are severed. Although it seems extraordinary now, it was widely performed in the middle of the last century as a treatment for mental disorder, especially manic depression, schizophrenia and bipolar disorder. Probably the most famous recipient (or rather, owing to the unsought circumstance of having a famous name, the most heartbreaking victim) of the procedure was Rosemary Kennedy, older sister of the future president John F. Kennedy, in 1941. The rationale for the procedure was that some mental disturbances were due to the malfunctioning of these nerve connections, such that messages passing across synapses from nerve centre to nerve centre kept getting stuck. The repetition of these messages, so it was thought, explained abnormal behaviour.

In Rosemary Kennedy's case, the behaviour took the form of extreme tantrums and violent seizures. Separated from her family and shut up in a convent, she became too violent for the nuns to control. Her father, Joseph P. Kennedy Snr, was worried by the taint of scandal and the implications for his sons' political ambitions, and authorised a lobotomy for his daughter, who was then aged twenty-three. It was a catastrophe. With the incisions in her frontal lobe, Rosemary was permanently damaged and unable ever after to speak more than a few words or to walk. From 1941 to the end of her long life (she died in 2005), she was in full-time care at St Coletta's School for Exceptional Children in Jefferson, Wisconsin.[4]

At the time, lobotomy was a new procedure that appeared to show promise. Its most zealous advocate, the Portuguese neurologist António Egas Moniz, was actually awarded the Nobel Prize in Physiology or Medicine in 1949 for his work in developing prefrontal lobotomy. It is commonly regarded as among the very worst decisions in the Nobel Foundation's history. Moniz's belief in this surgical procedure went way beyond the evidence. John Fulton, a historian of medicine at Yale, recounted a few years afterwards that Moniz's deduction about the efficacy of lobotomy was based on a talk that Fulton and an associate had delivered to an international neurological congress in 1935. Their report concerned the effects of removing the forepart of the frontal lobes of two chimpanzees. The procedure had turned the chimps from excitable to docile primates, and Moniz suggested, in a monograph, that the same effect might be produced in humans suffering from mental disorder. This was the intellectual inspiration for the thousands of lobotomies practised in the US and Europe in the following years, under a radical procedure that removed the entire orbitofrontal cortex. This is a part of the frontal lobe that is associated with decision-making. As Fulton drily and devastatingly stated: 'It soon became evident that, while the psychosis was often dramatically relieved, the patient, who might prior to his illness have been an individual of marked intellectual capacity, was reduced after the operation to a state of relative mental dullness.'[5]

The persistence of ECT

The disrepute into which the practice of lobotomy fell was far from the end for physical treatments for mental disorder and specifically for depression. Electroconvulsive therapy (ECT) has almost as bad a reputation as lobotomies, owing to its portrayal in popular culture – again, the figure of Jack Nicholson in *One Flew Over the Cuckoo's Nest* looms large. The procedure involves transmitting an electric current through the brain to induce a mild seizure. The method used to be that electrodes were placed on either side of the brain. More recent practice is to use only one electrode (which seems to limit the side effects of confusion in the patient). It is used for severe cases of depression or mania when talking therapies and medication have proved ineffective and for psychosis and schizophrenia. It's a momentous decision to deliberately pass electric shocks across a patient's brain, even though in modern practice it's done with anaesthetic and a muscle relaxant. The patient isn't awake and doesn't feel anything. Nor is there a physical convulsion of the body.

Does ECT work? Some psychiatrists say that for extreme cases it's effective, but it's not a unanimous view. One consultant and academic psychiatrist, Andrew Molodynski, is firm in his conclusion: 'It is by far the most powerful treatment for depression . . . As a treatment it is certainly not bad as long as it is done properly and for the right people, which is people with genuinely treatment-resistant depression.' Conversely an academic clinical psychologist, Richard Bentall, counters: 'My view is that ECT is a classic failure of evidence-based medicine. I don't believe that there are adequate clinical trials of ECT to establish its effectiveness.'[6]

If there is inadequate clinical evidence, then the task would in principle be to assemble it with randomised trials. But there's a problem with ECT: as both proponents and critics stress, there's an ethical objection to carrying out this procedure on people who aren't ill. NICE guidelines are that ECT should only be used as a last resort in cases of adult patients who are catatonic, or have prolonged or severe manic episodes, or have severe or moderate depression and their condition has been resistant to other treatments. As depression can be life-threatening, ECT is likely to be held in reserve as a possible treatment. It's generally given three times a week, and it has side effects. There's a risk of impaired memory. Against that, the risks of severe depression are very great indeed. It's possible for depression to be so severe that the patient loses the ability to function altogether.

As Mariam Alexander, an NHS consultant psychiatrist, has written of these extreme cases (which she says can respond positively to ECT): 'In liaison psychiatry, we see individuals so severely depressed that they have become catatonic – a state that means they may be unable to move, speak or eat. Admission to a medical ward is required in order to give nutrition via a nasogastric tube and medication to reduce the risk of dangerous blood clots forming due to immobility – seeing one of these patients, you'd be forgiven for thinking that the reason for admission was stroke rather than depression.'[7]

One of the difficulties in assessing ECT is common to other somatic treatments for depression. It's not clear how it works, if indeed it works at all. But a possible explanation is that it changes patterns of blood flow and the chemical make-up in the brain. An ECT Accreditation Service run by the Royal College of Psychiatrists reported in 2012–13 on the results of 1,895 courses of treatment conducted on 1,789 people in England and Wales. Almost all of the patients (1,712) reported an improvement.[8]

Physical treatments give way to medication

Applying physical 'curatives' to depression still exists, but is a minority treatment. While ECT is employed in some severe cases of depression (and there is statistical evidence that its use has slightly increased over the past twenty years), it bears a stigma owing to cultural depictions that are no longer accurate. Lobotomy has been held in public obloquy for a generation, and there it will undoubtedly remain. From the middle of the last century, tranquillisers and sedatives were prescribed in increasing quantities for dealing with a catch-all condition referred to as anxiety.

Far the most common way of counteracting depression is now with pills. Early medications for depression were successors to treatments by ECT and lobotomy. This pharmacological approach to depression gained momentum in the 1960s with the new generation of drugs.

The first drugs prescribed for depression were known as tricyclics (TCAs) and monoamine oxidase inhibitors (MAOIs). Their development was based on the theory that depression is caused by a chemical imbalance within the brain – that some chemicals, or neurotransmitters, are too low and that the drugs will prevent them from being absorbed by the brain. The process, as we have seen with SSRIs, happens like this. A nerve cell releases a neurotransmitter into a synapse. It's then either broken down or repackaged and taken back into the nerve cell, for use again. This repackaging is known as reuptake. The rationale of the antidepressant is that it blocks the reuptake (hence it's a 'reuptake inhibitor') and thereby makes more of the chemical available in the brain. This will in theory help stabilise the mood of a depressed person.[9]

The TCAs appear to work by blocking the reuptake of the chemicals serotonin and epinephrine. MAOIs work slightly differently. They block the effects of the natural enzyme monoamine oxidase. This enzyme breaks down serotonin and other neurotransmitters within the brain. So the net effect, if the problem is a chemical imbalance of neurotransmitters, should be the same.

There are problems with both these classes of antidepressants, though. They have side effects. The same is true of more recent medications, but there are specific side effects with these classes of drug. TCAs can affect heartbeats, causing them to become irregular. An overdose can lead to cardiac arrest. MAOIs can also cause an unnaturally high level of serotonin in the brain. This is a rare condition but it can raise blood pressure and cause kidney damage.

These classes of antidepressant have hence largely given way, at least as a drug of first choice, to newer ones. The reasons are part medical and part cultural. In the early 1960s the notion of anti-anxiety drugs (collectively known as benzodiazepines) attracted a sort of stigma that was then progressively dispelled. The Rolling Stones sang of 'Mother's Little Helper':

> You can tranquilise your mind
> So go running for the shelter of a mother's little helper
> And four help you through the night, help to minimise your plight.

The lyrics reflected the times. Anti-anxiety drugs were prescribed in vast numbers in the 1950s through to the 1970s, especially to women. These drugs are not in fact antidepressants but are often popularly assumed to be like them even though they work on different brain circuits. The brand names became a cliché of popular culture: Miltown, Valium and Librium. Where antidepressants were concerned, medical advances gradually produced new drugs that were not dangerous in overdose like the TCAs. The most common of the new drugs are SSRIs. That's what I took, but there are many of these apart from fluoxetine, which I was prescribed.

I've mentioned how SSRIs aim to block the reuptake of serotonin. A still newer class of antidepressant is known as serotonin and norepinephrine reuptake inhibitors (SNRIs), which block the reuptake of both serotonin and norepinephrine from the synapse to the nerve cell. Again, the idea is to address low levels of these chemicals. And then there are other antidepressants too. These include norepinephrine and dopamine reuptake inhibitors (NDRIs), which, as the name suggests, act by blocking the reuptake of norepinephrine and dopamine, thereby increasing their presence in the brain.

There are also cyclic antidepressants, not only tricyclics but also tetracyclics (depending on the number of rings in their chemical structure), which are a class of antidepressants that block norepinephrine reuptake and also block alpha adrenergic receptors. Finally, there are serotonin antagonist and reuptake inhibitors (SARIs), which, like SSRIs, block the reuptake of serotonin and prevent serotonin particles from binding with nerve receptors, but also redirect those particles to other receptors.

Recent thinking on antidepressants has moved towards the idea that they ultimately work not by blocking reuptake of neurotransmitters, whichever neurotransmitter it is, but by increasing neuroplasticity. This is the ability to form new connections between neurons. The theory is that depressed patients have impaired neuroplasticity, making it hard for their brains to move on from a negative mindset.

'Better than well'

Claims and counterclaims about the pharmacology of depression are strongly advanced. Some advocates of anti-depressants go well beyond asserting their strictly medical benefits. In the 1960s and 1970s, anti-establishment figures such as the psychiatrist R. D. Laing urged the benefits of psychedelic drugs. LSD (lysergic acid diethylamide), magic mushrooms and other hallucinogenic drugs were bound up with the cause of political revolt, but they were not a mere fashion accoutrement. According to one historian of the radical psychiatry of this era, hundreds of studies with psychedelic drugs took place, with research suggesting that they helped recovering alcoholics to abstain, soothed the anxieties of terminally ill patients, and eased the symptoms of psychiatric disorders such as obsessive-compulsive disorder.[10] And in literally a very small way, the market for psychedelic drugs has in recent years expanded once more. The practice of 'microdosing', or taking tiny amounts (around a twentieth of a recreational dose) of psychedelic drugs such as LSD and ketamine, is intended not to produce a hallucinogenic high but just to improve the way a person feels.[11]

Like the anti-anxiety pills alluded to in the Stones' lyrics a generation earlier, SSRIs also took on a cultural significance embodying the spirit of the age. Fluoxetine, whose trade name is Prozac, was approved by the Federal Drug Administration in 1987. In the 1990s, these pills were hailed by popular and also qualified authors as not just cures for depression or even for restoring sufferers to normality, but more broadly for enhancing wellbeing. For this type of drug, Prozac became a generic name, as recognisable in its own way as Kleenex, Hoover and Xerox. Peter Kramer, an American psychiatrist, extolled the virtues of SSRIs in a book published in 1993 titled *Listening to Prozac.* It's easy to caricature this enthusiasm as a reversion to the radical ethos of the 1960s, but in fact Kramer deals sensitively with the issue of potential side effects while noting correctly that Prozac has benefited millions of people.[12]

Kramer became famous for his thesis that Prozac doesn't merely aid recovery from depression but makes patients 'better than well'. It wasn't a deliberately provocative claim. He refrained from venturing into the subject again for many years, explaining only in 2016 to a specialist rather than popular audience: 'Regarding the science [described in the book]: I examined a phenomenon – I called it "better than well" – that I had encountered in my outpatient practice. Patients who had responded to treatment for major depression or obsessional states sometimes reported what they believed were separate medication effects on personality, largely a shift toward assertiveness and away from social anxiety.'[13]

Better than well: the phrase implied to sceptics an apparent abandonment by the medical profession of the virtues of rationality and self-restraint. And some advocates of antidepressants went way beyond even this formulation when discussing such wonder drugs as the new generation of SSRIs. They suggested that the mental disorders that Prozac was intended to treat were not illnesses so much as justified rebellions against modern social conditions.[14]

It's understandable that some would recoil at this rhetoric. The notion that mental disorder is not really illness but an alternative way of perceiving the world strikes at our conception of social order and convention. The arguments for 'listening to Prozac' suggest to the rationalist mind a wish to prettify the world rather than to reform or change it. They provoke popular suspicion of mind-altering drugs pushed by 'Big Pharma' on a docile and unsuspecting populace.

The backlash against antidepressants

There was, inevitably, a backlash to the arguments for antidepressants. Some of it was medically informed and much of it was not. The strongest challenge from within academia has come from Irving Kirsch, a Harvard psychologist. He takes issue with the entire notion that depression is caused by a chemical imbalance in the brain and argues that the benefits of antidepressants have been overstated and mischaracterised. In his book *The Emperor's New Drugs*, he states bluntly: 'The belief that anti-depressants can cure depression chemically is simply wrong.' Insofar as antidepressants are effective at all in countering depression, says Kirsch, 'most – if not all – of the effects of these medications are really placebo effects'. That's not to decry them. Kirsch acknowledges that placebo effects can provide substantial relief from depression, but argues that the chemical effect of the drugs may be small or non-existent.[15]

Kirsch's thesis has not drawn support from psychiatrists and psychologists. The literature is technical but the methodology matters. A position paper by the European Psychiatric Association demonstrates that the statistical analysis behind Kirsch's work is flawed.[16] When a popular US television programme ran with Kirsch's claims that there was no effective difference between antidepressant medications and placebos, the American Psychiatric Association condemned it as 'irresponsible and dangerous reporting'.[17]

Where Kirsch scores highly is in the popular writings of pundits rather than the research programmes of scientists. A prominent example is the British journalist Johann Hari. In a book titled *Lost Connections: Uncovering the Real Causes of Depression – and the Unexpected Solutions* (2018), he revealed that he'd taken antidepressants since his teens yet was now a critic of this medication. They hadn't seemed to work, despite increasing dosages.

Hari's epiphany came in discovering the work of Kirsch, whom he described as 'the Sherlock Holmes of chemical antidepressants – the man who has scrutinised the evidence about giving drugs to depressed and anxious people most closely in the world'. And, Hari argued: 'These drugs are having a positive effect for some people – but they clearly can't be the main solution for the majority of us, because we're still depressed even when we take them. At the moment, we offer depressed people a menu with only one option on it. I certainly don't want to take anything off the menu – but I realised, as I spent time with [Kirsch], that we would have to expand the menu.'[18]

Whereas scientists, who are driven by evidence, are cautious about our knowledge of mental disorder, Hari has apparently discovered the real causes of depression by talking to a few people he's carefully selected to yield the answer he wants. It's the same approach as those journalists who seek to cast doubt on the scientific consensus about climate change. The sole point he makes that is correct (namely that the chemical imbalance theory, which should more accurately be termed the reuptake theory, is unproven) is hardly novel, whereas his notion that only one option is provided to sufferers from depression demonstrates that he hasn't inquired very far about the state of health care. It's untrue. Clinical guidance in the UK states (emphasis added): 'If you have moderate or severe depression, you should be offered *both* an antidepressant and a psychological treatment – this should be either cognitive behavioural therapy (CBT) or interpersonal therapy.'[19] The NHS website states (emphasis added): 'Treatment for depression usually involves *a combination of* self-help, talking therapies and medicines.'[20]

To be fair, the same newspaper group that published Hari's claims also gave space online to Dean Burnett, a neuroscientist, to fairly summarise them: 'Hari may have the best intentions when it comes to addressing mental health problems like depression, but this doesn't seem like a good way to go about it. Asserting yourself as a maverick expert and backing your arguments up with suspect cherry picking of evidence and at-the-very-least exaggerated claims? Such a sensitive subject that affects millions surely requires a more thorough, thoughtful and specific approach than this?'[21]

It does; but so far as his maverick investigation goes, Hari picked an inadvertently revealing metaphor with Sherlock Holmes. Sir Arthur Conan Doyle's consulting detective is hardly the paragon of rationality he's generally made out to be.[22] Holmes's methods are simply wild speculations and guesses that, with remorseless serendipity, turn out to be always accurate. He advises Watson to maintain 'an absolutely blank mind, which is always an advantage'. This is not at all how scientists proceed: they have laws and theories, which are sets of logically consistent principles that explain a body of facts.

The clinical evidence

Even since Hari's book appeared, the largest international study to date has confirmed that antidepressants do indeed outperform placebos. This is a meta-analysis of 522 randomised controlled trials, comparing twenty-one different antidepressants of the 'second generation' (that is, since the early TCAs and MOAIs). More than 100 of these studies were previously unpublished. All of the antidepressants were found to be more effective than placebos in treating depression.[23]

The study was led by Dr Andrea Cipriani of Oxford University's Psychiatry Department, who says: 'Our study brings together the best available evidence to inform and guide doctors and patients in their treatment decisions. We found that the most commonly used antidepressants are more effective than placebo, with some more effective than others.' And he stresses, again contrary to Hari's caricature (emphasis added): 'Antidepressants can be an effective tool to treat major depression, but this does not necessarily mean that antidepressants should always be the first line of treatment. *Medication should always be considered alongside other options*, such as psychological therapies, where these are available.'[24]

This is an important study, not only for its conclusion about the efficacy of antidepressants but also as a model of how science works. It assembles evidence, assesses it and in drawing contingent conclusions doesn't go beyond it. All the data has been made publicly available and put online. That's crucial, so that the results can be assessed and replicated by other competent researchers. The limitations of the study are also clearly stated. These are scrupulously summarised in an analysis on the NHS website. The results are for use of antidepressants after eight weeks, so the study doesn't assess the long-term use of these drugs. The trials included in the study were of variable quality and did not report on the side effects of using antidepressants or withdrawal symptoms from discontinuing with them. And the study did not assess how antidepressants performed in combination with psychological treatments or as an alternative to them.[25]

Side effects and risks

The data suggests that among younger patients – children, adolescents and young adults – there is a slight increase in suicidal thoughts when on a course of antidepressants. This side effect has been publicised, especially in the case of an SSRI antidepressant called paroxetine. The drug's trade name is Seroxat in the UK and Paxil in the US. A study by its manufacturer, GlaxoSmithKline, showed an increase in emotional side effects, including suicidal thoughts, among children taking the drug compared with those taking a placebo. That doesn't demonstrate a causal link, nor has it prompted psychiatrists to stop prescribing it, but media coverage of the finding has surely contributed to mistrust of SSRIs.[26] This risk of suicidal thoughts among young people (GlaxoSmithKline stressed its study had found no cases of actual suicides) underlines that it's vital that these drugs be dispensed only under medical supervision. Depression needs to be treated rather than hoping it will go away. As far I was concerned, I had thoughts of death and suicide often during depression, but because of the depression and not the medication to deal with it. The drugs damped down my depressive thoughts rather than stimulating them. That's how they're supposed to work. And if, indeed, some pundits have overstated the transformative effect of antidepressants, there is a reassuring implication here too: these drugs are not easy to abuse, in the way that legal 'highs' can be taken to dangerous and deadly excess. Not at all.[27]

Whereas antidepressants can stabilise the moods of a

depressed person, they won't have an effect on someone who is not suffering from depression. They act on depressive illness, rather than on low mood generally. For a normal person they may induce a slight sluggishness, but that's the sum of their effects; they won't act as a stimulant. There is no secondary market in antidepressants, for that reason, nor are they sold on the street from dealers, as would inevitably happen if they produced feelings of elation and happiness.

There is no danger, either, that users will become addicted to antidepressants as they might even to sedatives, let alone more destructive drugs. I have a colleague who has taken antidepressants for years, to the extent that his daughter refers to them as 'Daddy's mad pills'. He takes antidepressants day after day, year in and year out, not because he's in any sense hooked on them but because he suffers from continuing depression and the drugs ease his symptoms. That's a good thing, not a criticism. I don't want to overstate the point. There is some evidence that antidepressants can lose efficacy for an individual over time, and someone who gains benefit from them once may not do so when they are prescribed for another episode. And while antidepressants aren't addictive, there can be profound and unpleasant withdrawal symptoms (which have been observed especially with paroxetine and venlafaxine). But broadly, they are beneficial in the treatment of depression.

Another factor that militates against the abuse of antidepressants is that they take a long time to work. The usual period after which they become effective is between two and four weeks. I didn't even notice an improvement at that interval and feared they were having no effect at all. This is quite unlike the experience of recreational drugs, where the user has a short-term rush and, insofar as they reflect on it, wishes for no long-term effect. The side effects of antidepressants – including, for some people nausea, headaches and vomiting – tend to act quicker than the benefits. There are hence no obvious attractions for the casual user in this class of drug, and there are many deterrents to taking it.

Finally, there is no typical patient for whom antidepressants work. For some people suffering from depression the drugs are clearly effective, but this is not true for all. The variety of medications may cause scrupulous doctors to prescribe several before settling on the most effective, and the most effective dosage varies greatly from patient to patient. If the dose is below a certain therapeutic threshold, it may accomplish nothing. If the dose is sufficient but the patient abandons it too soon, antidepressants may likewise fail to accomplish anything.

In my case, I wondered whether the drugs were having any beneficial effect. I felt terrible and knew there were physiological effects on me that I didn't want. I was advised what these might be, and at least two did eventuate. I had dreams that were vivid, recurring and sometimes disturbing, but attributed no symbolic significance to them. I have no doubt this was the most prudent course and it saved me a lot of time not to scrutinise them and wonder what they meant. And, already torpid and exhausted through grief, I lost any trace of physical desire. In his final years Tony Benn remarked of himself that 'as you get older, all desire goes – medical conditions help with that – so I think of myself now as a biological Buddhist'.[28] Medication did the same to me. It was of no practical consequence, for I was unattached and no one, I imagined, could have desired me in return. But it was a fact. If Charlize Theron herself had been dispensing my prescription, I'd not have noticed.

For doctors who aren't conversant with psychiatric disciplines, there must be an incentive – if only out of risk-aversion – to curb the antidepressant medication rather than prescribe an increase in the dosage. Yet this response is misguided. Antidepressants won't have an effect in the short term; they may well have unpleasant side effects, but any benefit they have will take weeks to be realised. It's the same with any chronic illness, like diabetes or heart disease. If you throw away the medication, you'll dispense with the chances of recovery; and if you continue with it, at best the symptoms will be eased rather than the underlying condition cured.

'Handed out "like sweeties"'

Though the case against antidepressants is weak, and the dangers much overstated, there remains a stubborn argument in public debate that doctors prescribe them too readily. The evidence is that they are more effective with moderate to severe cases of depression than with mild depression. It's undoubtedly possible in theory that doctors are prescribing the drugs to people who won't benefit from them – who in fact aren't clinically depressed but are suffering natural responses of grief or sadness to events that disrupt their lives. You'll find plenty of media coverage of this accusation, though the journalists who write this copy never state what the 'correct' level of prescriptions should be.[29]

It's an empirical question. And the evidence is far from conclusive that doctors are doing what the critics accuse them of. A rise in antidepressant prescriptions doesn't necessarily mean that more patients are taking the pills. It may equally be not only that the same patients are being prescribed higher dosages (which is what happened in my case) but also that their treatments are longer, lest they have a relapse by stopping too early, and that medical practice is thereby catching up. The notion that antidepressants are being carelessly prescribed is purely anecdotal and serves to reinforce the stigma against mental illness. Indeed, studies point to the opposite conclusion, that depression is under-diagnosed and under-treated. The late Ian Reid, a Professor of Psychiatry at Aberdeen University, was dismissive of such jaundiced media coverage: 'The idea persists that GPs are handing out antidepressants "like sweeties". We screened nearly 1,000 general practice attenders in Grampian for depression and scrutinised the prescription decisions made by thirty-three GPs. Almost half of the depressed patients we identified were unrecognised and, contrary to popular stereotype, GPs were cautious and conservative in their prescribing for those that they did diagnose.'[30]

Even if the evidence did suggest that the clinical definition of depression had become too loose and that antidepressants were being overprescribed, this would still be a far cry from the notion that major depressive disorder is mythical. Yet in fact antidepressants are not doled out as a crutch to the emotionally needy or as an alternative to delving into a complex emotional state. They offer a route towards better treatments for the causes of the destructive thinking that can cause a depressive state. They were, in that respect, the beginning of my own recovery.

My experience of antidepressants

That's where the science stands on antidepressant medications, as I write. The evidence is that they work – not for everyone, and not to the same extent, but people with depression are more likely to experience an improvement in their symptoms if they take antidepressants than if they take a placebo. And this is how they worked for me.

Once my severe clinical depression had been diagnosed by Dr Nathan, he put me on a course of antidepressants. These pills could be obtained right away, so he explained what they were and how they worked. Contrary to the critics of antidepressants, Dr Nathan said the pills were aids to recovery that clinical trials had validated. They made patients' lives better. I readily assented to anything that might help, but in this case it was a soundly based leap of faith.

Dr Nathan prescribed Prozac. He also prescribed for me a sedative and hypnotic called Zolpidem, to help me sleep. He had many points of advice about starting a course of antidepressants but just one about the sedatives. It's easy, he said, to become dependent on sleeping pills, and this was the very last thing I needed in my current state. He advised taking them for three nights in succession to try and get into the habit of sleep, and then to put them aside at least every other day, if not for a few days. We parted with, at last, a glimmer of hope on my part that I now knew, with a medical diagnosis, what had been going wrong, and that the frustrations of my sessions with the therapist were not an abject failure on my part. I went to the pharmacy to get the pills, believing that I could at least have an uninterrupted night's sleep, and waited for a call from a psychological practice that Dr Nathan had referred me to.

Having had the diagnosis and received the prescription, I felt momentarily brighter, which was just as well. My mental state had not improved, and indeed could hardly have been worse. It was becoming still more noticeable to others too. Tearfulness could come at any moment, whether or not I was ruminating. Just being in a crowded place would render me anxious and desperate to escape. In the office, one of my close colleagues would sometimes playfully bat me round the back of the head if he thought I was engrossed in thought. It was affectionate and not in the least combative. Sitting at my desk, I knew beyond any doubt that I'd lose composure and weep if he did it this time. He must have sensed that I shrank, and refrained that day and has done forever since. Underlying everything was the constant guilt and mental flagellation. On the day I collected the pills, I came home from the office having accomplished nothing but with the hope that there was a way forward. I went to bed in the middle of the evening, having taken an initial dose of the fluoxetine and a sleeping pill. I couldn't sleep. And I wasn't calmed by the antidepressant.

I shouldn't have been surprised. These were not instant expedients, let alone panaceas. It was a bad move to regard them as a turning point, as I was bound to be disappointed. On a dark day it's inevitable you'll want to curtail it by retreating to bed and hiding as soon as you get home. All this accomplishes, however, is to prolong the agony by ensuring that the travails of the day remain in your mind.

I found a compromise in the end, by sitting up even on the worst of days till 10 p.m. and only then – and only if the terrors hadn't diminished – taking the sleeping pills, regardless of the risk of excessive consumption, along with the fluoxetine. It took a long time but, confounding my expectations, the pharmacological route was effective. With these powerful prescription drugs, the antidepressants and sedatives, my wilder imaginings began to stabilise. It was an immense temptation to take more of them, but in the case of the sleeping pills I acknowledged and generally adhered to what Dr Nathan had stressed: I couldn't emotionally or mentally afford to get hooked on them. They were a relief when I was able to take them. Sometimes they didn't work; more usually they did, and I had a brief sense, lasting a mere instant when I awoke around 5 or 5.30 a.m., of being rested – whereupon the depression crowded back in once more.

We know how sedatives work. They're not sophisticated science nor do they alter the person's psychological state.

They aren't just taken by people who have mental disorders. They have an effect on most people. If you're feeling in a hyperventilating mood, then sedatives will calm you down and lessen your tension. If you're sluggish but your mind is disturbed, which was the case with me and my depression, a sedative can help knock you out and get you to sleep. It's a humane recourse that has to be temporary but will usually work.

As for the antidepressants, I had long been sceptical about their widespread use while knowing next to nothing about them. As with ECT, our instincts are shaped not by the science but by the portrayal of this treatment in popular culture and by its reputation. Christopher Hitchens caustically referred to antidepressants in his memoir *Hitch-22* (published in 2010, immediately before he knew of his advanced cancer):

> Perception modifies reality: when I abandoned the smoking habit of more than three decades I was given a supposedly helpful pill called Wellbutrin. But as soon as I discovered that this was the brand name for an antidepressant, I tossed the bottle away. There may be successful methods for overcoming the blues, but for me they cannot include a capsule that says: 'Fool yourself into happiness, while pretending not to do so.' I should actually want my mind to be strong enough to circumvent such a trick.[31]

Yes, I should want to as well, and that's how I reasoned; but it's not like that. The capsules aren't a trick. Christopher might as well have said that he'd like his physical immunities to be strong enough to withstand disease. Illness is not a test of willpower, to be conquered by an assertion of will, but a matter of fact. It is very important that depression get treated. If it is not, then needless suffering will ensue, with a risk of suicide and a toll on physical health and the quality of work and relationships. Christopher may have been thinking – he almost certainly was thinking – of the soma drug in Aldous Huxley's *Brave New World* (1932), which is officially distributed to produce a popular happiness: 'By this time the soma had begun to work. Eyes shone, cheeks were flushed, the inner light of universal benevolence broke out on every face in happy, friendly smiles.'[32]

Antidepressants aren't simulations of happiness. They are medicines designed to alleviate real and distressing symptoms. They are one part of the Hippocratic impulse to diminish suffering by alleviating devastating illness. I've no doubt that antidepressants served me well and aided my recovery. They solved nothing in themselves, but I navigated problems and coped with the daily decisions of life more easily merely by having a greater predictability of my emotional state.

Not 'better than well', but easier to get better

That's the unspectacular case for antidepressants. They don't make you well. They don't make you better than well. They make it easier for you to get better. They enable you to find the emotional space in which to train yourself to get well.

The sensible and cautious response of sceptics to any more expansive claims for antidepressants is that our knowledge of how they work is limited. We don't know what causes depression or what changes happen in the brain at the same time. The idea that it's due to a chemical imbalance in the brain is a very common view, but it's not universally held and the evidence for it is inconclusive. Though there's plenty of evidence that antidepressants improve the quality of life of those who suffer depression, we don't properly understand how and we know they can have side effects.

Ignorance is not the only reason for being wary of inflated claims for the benefits of antidepressants. Above all there's the fact that they do not shorten the lifespan of depression. They are not a cure. What they do is alleviate the symptoms. It's the same with medication for the common cold: there's no cure for colds but there are drugs that, for most sufferers, will ease headaches and sore throats, open up blocked sinuses and aid sleep. Antidepressants don't dispel depression: they stabilise the sufferer's mood, while the passage of time or human intervention gets to work.

The usual practice with antidepressants is for the doctor to prescribe a low dosage at first and then to monitor the patient's mood to assess whether the drug is working. I initially took one fluoxetine pill, of 20 mg, at night, swallowed with water. They were green and white capsules that for me became a distinctive and comforting image. The dosage initially prescribed by Dr Nathan didn't seem to be effective so he put the prescription up to 40 mg. Then it kicked in. I didn't myself sense any obvious change in mood, but it was noticeable to others. One of my close friends and colleagues, who had seen me at my very worst, merely said good morning one day and was surprised when I returned the greeting. Apparently, it was the first time for months that I'd shown awareness of my surroundings. The drugs worked in the sense not that I was better but that I could handle the days better.

7

PSYCHOLOGICAL TREATMENT

> The sensation of falling off
> the round, turning world
> into cold, blue-black space.
>
> Elizabeth Bishop, 'In the Waiting Room', 1976

The clinical evidence for the effectiveness of SSRIs is strong. The other principal movement in treating mental disorder over the twentieth century stressed therapy rather than surgery. And here the case is mixed. One important branch of therapeutic intervention has for good reason been largely superseded. This is the tradition of the most famous name in the field, Sigmund Freud, whose theories, while not principally concerned with depression, are relevant to it.

Freud hypothesised that some patients become depressed because of an earlier experience of parental rejection or loss. A Freudian approach would see depression as an expression of anger towards the lost parent. As the parent is no longer around (and might indeed have passed away since that early experience), the patient's anger is directed inwards towards the self rather than outwards. The loss of a person in one's life (especially in the context of the break-up of a marriage or important relationship), or loss in the less tangible sense of a thwarted ambition, can cause depression by forcing the sufferer to experience once more that formative experience of parental rejection.

Critics of Freud are legion, and their unassailable point is that these theories are very hard to test. The problem, they argue, is not so much that Freud's explanations are wrong as that they bypass scientific inquiry altogether. Frederick Crews, once an exponent of a psychoanalytic approach in his field as a literary scholar, is a leading critic of Freud, and has cogently argued just how limited is the analytical and predictive power of psychoanalysis: '"Psychoanalytic method" – the analysis of (allegedly) free associations, of dreams and slips, and of the "transference" – is much the same as it was a hundred years ago, and it is helpless against the contaminating effect of suggestion. That is why we see so many warring psychoanalytic schools, each boasting "clinical validation" of its tenets.'[1]

Yet these ideas did prove influential in a movement, within the field of mental health, towards psychodynamic psychotherapy in the 1960s. Mary Cregan is an American literary scholar who suffered severe depression, and attempted suicide, after the death of her two-day-old baby. In a fine memoir she notes that the American Psychiatric Association's first two editions of the *DSM* (in 1952 and 1968) enshrined the intellectual legacy of Freud. She writes: 'The APA also recommended that all psychiatrists should be trained in Freudian and psychodynamic concepts, and psychoanalytic training institutes began to spring up in major cities. Psychoanalysts chaired the psychiatry departments at most universities and held the medical directorships of most psychiatric hospitals, including the one where I would be treated.'[2] The afterlife of these ideas was what I encountered in my fruitless meetings with the therapist. I came away from them bewildered, rudderless and appreciably worse than when I'd gone in.

Freud and Kraepelin

Though his name is far less well known to the public, Freud's compatriot and contemporary Emil Kraepelin was ultimately to prove more influential in treatments for depression. Kraepelin was a German professor of psychiatry who theorised that psychiatric disorders were the result of biological conditions. The relevance of his approach to modern studies of depression is that he distinguished between two principal forms of psychosis.

These were schizophrenia (which he termed 'dementia praecox') and manic depression. The first was, Kraepelin theorised, a neurodegenerative disease like dementia. (Alois Alzheimer was a student of Kraepelin's at Munich University, and Kraepelin coined the term 'Alzheimer's disease' to refer to his discovery of presenile dementia.) It was hence untreatable, leading to a progressive and irreversible loss of cognitive faculties. Depressive disorders, by contrast, were episodic and did not cause brain damage. They could be treated and the sufferer could recover.

The term 'schizophrenia' was coined by the Swiss psychiatrist Eugen Bleuler. He discussed manic depressive psychosis as the other long-lasting mental illness, which would include severe depression as we think of it now. Whether the distinction between these two forms of illness is quite as rigid as Kraepelin and Bleuler theorised has been queried by psychiatrists, but it forms an essential diagnostic tool in the history of treating mental disorders. The notion that depression may have a biological basis challenges the psychoanalytic approach. In psychoanalysis, the therapist helps the patient delve into sublimated memories and deep-seated mental psychoses. By this means can depressive disorders be not just pacified but healed.

Kraepelin himself judged that psychoanalysis did not have a firm scientific foundation. My experience of its modern successors was not happy, and I have since wondered why. Here are my answers, born of that treatment.

Entering psychotherapy

I confided in a handful of close friends when I started seeing the therapist. Only one of them was enthusiastic and believed I would find the answer to my problems. Others were relieved, not because they had faith that I would be cured that way but because I was frightening them with the stuff I was saying and the things I was doing. Another friend, a close colleague and a leading political columnist, was intuitively sceptical, and with good reason as it turned out.

He asked, in genuine interest about the treatment and out of concern for my recovery, whether the sessions involved me lying back on a couch. I was able to tell him that they didn't and that the stuff of countless *New Yorker* cartoons and caricatures was wrong. In fact I sat in a chair facing the therapist in the rather shabby consulting room in her flat. It was far removed from the stereotype of intensive interrogation from the therapist's side and free-form association of ideas from mine.

The 'psychodynamic' form of therapy I underwent was not the full-blown psychoanalysis practised by Freud and his followers, but it had that intellectual lineage. Its premise was that psychological factors rather than chemical states were at the root of my depression. And the therapist was convinced that by examining formative and crucial episodes in my life, the key to my present unhappiness and mental turmoil would be found.

There was a method to this. The therapist listened – indeed, that's overwhelmingly the sum of what she did – rather than hectored. The purpose of therapy is never to tell the patient the answers but to try and arrive at a shared understanding of what the problems are and how they might be resolved. Our initial session was geared towards discovering the events that had tipped me into serious illness and the issues we should focus on in future. If it had worked, we would have progressively narrowed down the impediments to my normal state and identified what was preventing me from forming normal relationships – not only in the romantic sense, but with family, colleagues and friends.

What went wrong? I'm certain that the therapist judged me from the outset as a chippy and obstreperous personality who, even if not by design, would cause trouble. I can't much blame her. I kept going to the sessions despite my mounting frustration that the therapy was not helping and that the failure of it was intensifying my mental turmoil. It got to the stage where I just talked about my memories, distant and near, regardless of any applicability to anything, let alone my depression, because that's what she seemed to expect. I must have seemed an ungrateful and intractable case. But for all our mutual incomprehension, and my own character flaws and idiosyncrasies, I'm convinced there's a wider explanation for the failure of our sessions together. It's that dynamic psychotherapy itself was of no use to me in countering depression. That doesn't mean it's wrong or damaging. For some people, it will provide comfort and aid in navigating difficulties at any stage in the voyage of life. But mental illness is something else. I don't believe that dynamic psychotherapy is well fitted to deal with it. My reason is that the evidence doesn't support its theory of why depression occurs or show that it works with depressive disorder. If it did, I'd recommend it. As it is, there are potential pitfalls that scrupulous psychotherapists will be aware of but others may be susceptible to.

Psychotherapy, psychiatry and psychology

Let me explain these differences in function. In health care there is, as we've seen, an artificial and unhelpful division between the specialisms of caring for body and mind. The artificial distinctions go wider than this and are enshrined within mental-health care too. There is a strict taxonomy of who does what. A psychiatrist is a trained doctor who deals with mental disorders and can prescribe medication. A clinical psychologist deals with mental disorder (and a lot more, but this is the field I'm concentrating on here) from a behavioural rather than a medical standpoint and intervenes with 'talking therapies' to change a person's habits of thought. There's a plausible case that this distinction is too rigid and that a psychologist should be able to prescribe antidepressants; as it is, a psychologist dealing with a case of moderate to severe depression will invariably work closely with a psychiatrist or the patient's GP. Finally, a psychotherapist is a much looser term than either a psychiatrist or a psychologist, to describe anyone who treats people for emotional problems.

But here's an oddity. A psychiatrist is a qualified medical doctor who uses a variety of treatments for mental illness. To be a clinical psychologist you need a degree accredited by the British Psychological Society, three years of postgraduate study leading to a doctorate, and at least a year of clinical work experience. What, therefore, is required to qualify as a psychotherapist? The answer is nothing. Literally nothing at all. Anyone can call themselves a psychotherapist or, looser still, a counsellor. There are umbrella bodies that act as professional associations in the field, notably the British Association for Counselling and Psychotherapy (BACP), and offer training courses and their own diplomas and certificates. But as the BACP disarmingly says on its website: 'BACP, and other professional associations, set their own standards for training in counselling and psychotherapy as there are no compulsory training courses or qualifications for therapists.'[3]

Do you find that strange? I did when I started my inquiries into depression, and my puzzlement doesn't diminish. Consider that if you want to set yourself up in business to offer financial planning, accounting services or legal advice, there are qualifications you are required to have earned. Otherwise you're not a financial planner, accountant or lawyer. You may have extensive general knowledge and interest in the fields, but you're not competent to dispense advice and no one who takes it will be legally protected. The same is true of many other professional fields, including, most obviously, medical advice. When a doctor prescribes treatment or medicines, the health and sometimes the life of the patient are at stake. But for mental health, this is not the case; anyone can do it.

The dangers of unregulated therapy

It doesn't necessarily follow that helping people cope with depression should be a regulated activity. After all, the simple human impulse to help and to listen to those in distress may in itself be of solace to a depressive and can help them feel valued amid the despair. But that's what it is. It's not medical treatment, nor is there science behind it. And the ease with which it's possible to describe yourself as a therapist can have direct and very human consequences for the patient. Here's an illustrative news report.[4]

> Unaccredited online therapists are 'preying' on the desperation of people with mental health problems as the NHS struggles to meet rising demand, in many cases exacerbating people's issues, experts warn.
>
> Vulnerable people are being exploited by 'unethical' private websites which charge large sums of money for therapy sessions via online chats – with some services even being used as a tool to project religious and spiritual beliefs.
>
> There is growing concern many of these websites operate using unaccredited counsellors who are not medical professionals and do not have the minimum training standards to treat serious mental illness.

This is exploitative and scandalous. Yet it turns out that the 'experts' cited in the report unintentionally demonstrate the problem that the story is highlighting. A spokesperson for the BACP is quoted expressing her concerns. She points out: 'There's no regulation, except for small professional organisations that insist on members having been trained. But those are specialist organisations. Anyone can set up a website that charges for therapy.'

I don't mean to elide the difference between well-intended therapy and religious proselytising. It matters, and the proselytising is the most disturbing aspect of this story. I've come across this type of meddling in people's lives. There is a small subculture of evangelical Christianity that regards mental illness as the fault of the sufferer and insists that the remedy is biblical counselling.

This sort of intervention in the field of mental health tests the limits of a society that observes the Jeffersonian principle of religious liberty. A ministry that blames people who are sick for their own supposed moral failings is a guarantee of their continuing mental torment and its attendant dangers. To a person in a depressive state who is already convinced of their own depravity, that message will intensify a sense of guilt and hopelessness. To the psychotherapy profession faced with religious obscurantism and secular hucksterism, the conclusion is that they need to provide accredited courses, the completion of which will allow practitioners to put letters after their name and thereby give confidence to their clients. Spreading awareness of this self-regulatory system would, they maintain, deter 'unqualified' people from offering online or personal support to people who need help. That's the message of the BACP when faced with such exploitative behaviour.

I don't doubt the therapists' good intentions, but to complain about unqualified people in this context is begging the question. It presupposes that people who are trained in the techniques of psychodynamic psychotherapy have a real function in dealing with depressive disorder. The NICE guidelines are suitably cautious. They advise that, if you have depression, 'you may be offered counselling or short-term psychodynamic psychotherapy . . . However, your healthcare professional should explain that it is uncertain whether counselling or short-term psychodynamic psychotherapy are helpful for people with depression.'[5]

Judgement on psychotherapy

Psychotherapy has a role but also a limit. It has many subdivisions, for example between psychodynamic, behavioural and cognitive-behavioural approaches, and over brief or extended periods, and for individuals, couples, families or groups. My particular concern, born of experience, is with the role of psychodynamic therapy, which is a variant informed in part at least by the ideas of Freud and psychoanalysis. Its practitioners are with few exceptions good-hearted people who genuinely wish their clients a resolution of their inner conflicts and of their difficulties in relationships. These psychotherapists are not charlatans, let alone religiously motivated mountebanks; they study hard and have generous impulses. They may have put in a great deal of training on their various accredited courses. There may be value in their approach for some mental health problems: there's scant evidence for its effectiveness in treating mood, anxiety and psychotic disorders, but it does inform treatment of personality disorders, eating disorders, forensic psychiatry and the practice of group and family therapy.[6] Yet their techniques for dealing with depressive disorder are quite speculative.

In dealing with depression the methods of psychodynamic psychotherapy presuppose that the disorder can be solved by examining its deep-rootedness in our psyche. A mutual exploration, driven by the client but with the encouragement and gentle navigation of the therapist, will uncover the origins, and even if it doesn't the journey itself will be fulfilling. That at least is the theory behind it. I don't believe I've caricatured this approach or the conviction of these psychotherapists that their work is more effective and enduring, dealing with causes rather than merely ameliorating symptoms, than the more circumscribed psychological treatments that the clinical guidelines support.

To the extent that dynamic psychotherapy is helpful, however, it is likely to be with people who are not really ill: perhaps those who have undergone family difficulties or a relationship break-up, or have simply reached a stage in life that is different from what they'd expected or intended. Talking to a stranger who will not judge, who is personally compassionate and who will make all the time available that you need may help those, whether couples or families or individuals, who are finding life a struggle. And all of us at some time do. There are even predictable stages of life where this is commonplace: recall Thomas Cole's *Voyage of Life*, as his subject in maturity drifts between forbidding rocks and withered trees and prays for deliverance.

Psychotherapy helps in these cases not by finding and extirpating the causes of depression but by doing something much more prosaic. It can bring a kind and sympathetic person into the lives of those who have problems, sadnesses and disappointments. Sometimes we need the kindness of strangers, and if they have an air of wisdom and authority, then so much the better for the solace they provide. In *The Mill on the Floss*, George Eliot matchlessly describes such a calming presence when Maggie Tulliver needs it:[7]

> She felt a childlike, instinctive relief from the sense of uneasiness in this exertion, when she saw it was Dr Kenn's face that was looking at her; that plain, middle-aged face, with a grave, penetrating kindness in it, seeming to tell of a human being who had reached a firm, safe strand, but was looking with helpful pity toward the strugglers still tossed by the waves, had an effect on Maggie at this moment which was afterward remembered by her as if it had been a promise. The middle-aged, who have lived through their strongest emotions, but are yet in the time when memory is still half passionate and not merely contemplative, should surely be a sort of natural priesthood, whom life has disciplined and consecrated to be the refuge and rescue of early stumblers and victims of self-despair.

To be with someone of inherently generous impulses, and with a natural authority born of experience of personal travails, is itself therapeutic. It calms by giving a sense of sympathy and an assurance that others have weathered life's tempests, however hard the present may seem. It's part of the voyage of life too.

But, where clinical illness is concerned, this is like saying that a dispenser of homeopathic remedies can have a calming effect on a patient merely by having a sympathetic manner. It may describe what's happening, but the treatment is not useful to someone who is ill. In that case, what's required is a medical doctor and treatments that are effective. The same principle holds with mental health: if you're seriously ill, you'll want someone who is sympathetic but above all you need someone who is trained in scientific methods based on clinical evidence. Perhaps if you're suffering from severe depression you could find a psychotherapist who is aware of their own limitations and will work in concert with psychiatrists, psychologists and GPs who prescribe antidepressants. But I didn't get it, and encountered conflicts of interest in psychotherapy once it was thrown into this mixing of specialisms.

The psychotherapist was the first professional I saw when I realised there was a serious problem. I had little notion of what I'd find, but I knew what I wished to get out of our meetings: recovery. I reasoned that if I saw different types of specialist in addition to the therapist I'd been referred to, the chances of dispelling the agonies and emerging once more into normality would increase. So I explained my plan to the therapist at our second session. To my consternation, she advised against it. Yes, seeing a doctor was sensible and is what she'd suggested I do in the first place – but not someone else whose field was talking therapies. Oh, no: that was not a good idea at all. She asked me briskly what I thought I'd get out of it. I explained the scheme that I'd carefully thought up. My first requirement was not to feel absolutely bloody and horrific; I assumed that seeing someone who could explain how to get better would be urgently needed relief. Seeing the psychotherapist in addition would help guard against a recurrence and secure my long-term mental health.

My scheme seemed to me then to make sense, even though I now know it didn't. The therapist objected to it for a different reason. She insisted that for me to see someone else would work at cross-purposes to what we were trying to achieve. She advised that this would be like putting a sticking plaster on a wound, when what we needed to do was to identify the problems that had caused the wound to erupt in the first place. In her firm professional opinion, I was stricken in this state and to go for a short-term fix would be ultimately futile and a missed opportunity. Instead, I needed to be mentally healed and our mutual exploration of themes in my life would enable us to uncover what had caused my depression.

I don't recall her exact words, at this distance, but this was her argument and it was sincerely meant. It was terrible advice, as it dissuaded me from seeking the treatment that could genuinely have been of benefit and that did eventually bring me to a full recovery. I can't regard the therapist's urgings as any breach of professional ethics; I'm certain they weren't. She was trying, by her criteria and according to whatever training she had, to guide me into a course that she believed would have lasting benefits for me rather than be a temporary palliative or a placebo. And, for her, the realm of disturbance was the unconscious; by delving into it and allowing me to express what I found there, we could cure my illness.

The conflict of interest in psychotherapy

My therapist's conduct was irresponsible. The blame for that lies not with her but with the approach she was sincerely, devotedly, intent on pursuing. The fault lies with inflated claims for dynamic psychotherapy itself. The clinical evidence alluded to by the NICE guidelines does not find that this form of psychotherapy is effective as a method of tackling depression. And in my case it undermined, rather than – as I'd naively imagined – cooperated with, real mental-health disciplines. There is a potential conflict of interest that psychotherapists have when dealing with sufferers from depression and other mental disorders, and that they need to guard against. Whereas psychiatry, pharmacology and psychology are all disciplines concerned with making mentally ill patients *better*, dynamic psychotherapists have an interest in keeping their clients *talking*. Here's an example from another newspaper report that I read in the year of my depression:[8]

> Having spent six months overcoming a period of depression in private therapy sessions, Gemma felt ready to stop. But when she raised the issue with her therapist, she met resistance.
>
> 'For two months, I told her that we need to cut down or stop altogether,' she says, 'and every time it would be diverted into a discussion about why I'm not willing to spend £60 a week on myself.' The absurdity of the situation reminded Gemma of trying to cancel her contract with Sky. She says: 'I contacted them eight times. They'd tell me that changing to BT would be a bad idea; every time, I ended up saying that I'd "have a think".'

There is, in this story, further worrying evidence of clients being badgered by therapists ('he called me for three days, telling me I wasn't in the right mind to make such a decision'). Again, you can hope that these therapists' methods of persuasion are atypical. The article goes on to quote a lecturer in counselling who says that the client must be in control of the process, and I'm certain that most therapists and counsellors are scrupulous in observing this maxim. That's not the problem I have with these accounts. It's instead about procedure: there is no natural end point to the treatment that the psychotherapist provides. That's why Gemma's therapist pressed her to continue. The therapist perhaps reasoned to herself that she, rather than Gemma, was in the best position to judge the state of Gemma's progress. What's to stop her? The profession of psychotherapy and counselling will collectively say that the wishes of the patient are paramount, but that isn't an adequate protection against a therapist who sincerely believes that the sessions should continue. In principle, the sessions could go on and on, never reaching a conclusion. Gemma had the independence of mind to judge that the sessions weren't of value to her and the strength of character to make her own decision to stop them. Had she not done so, then the therapist would have persisted.

That's where a conflict of interest may lie between the client and the therapist. It's not merely, or even at all, the financial incentive. When I recounted to my friends that I had taken the decision not to see my psychotherapist again, they were incredulous that on our first meeting she'd proposed, and I'd agreed to, three sessions a week over the next few months. It would have mounted up. The payment per session was £75, so the weekly fee was £225. But if these sessions had been valuable, it would have been money well spent. Rather, the problem may be the perverse incentive whereby the therapist wants to keep on talking, and getting the client to talk. This is not the same ethos as that of a medical professional or a psychologist. In those disciplines, the practitioner wants to see the back of the patient after successfully treating them in order to help others who are ill. That's the point of the consultation and the treatment.

There's nothing wrong with a professional relationship in which a person wishes to talk about life's problems with a sympathetic stranger, and to pay for the service. But that's not the same as alleviating or curing illness. On the contrary, because conversation is limitless, so the psychotherapeutic experience is without natural bounds. There is no stage at which the therapist will say: 'That's it. You're better. I've taught you techniques that you can use to scrutinise your mental states. There's no more for us to do except converse, as you would with any other sympathetic stranger. But you've got a life to lead and I've got people to see who need me.'

If you want to talk open-endedly with a psychotherapist about life, and you get on with them, don't hesitate to go ahead. It may add an extra dimension to your thinking and even benefit you. There are always some things we are diffident about revealing in our personal lives even to friends. Talking about them in the confidentiality and welcome of a counselling service is an option.

But this was not for me a route to dealing with mental illness. My experience was instead similar to the way Aaron Beck has caustically described the psychotherapeutic approach:

> The troubled person is led to believe that he can't help himself and must seek out a professional healer when confronted with distress related to everyday problems of living. His confidence in the 'obvious' techniques he has customarily used in solving his problems is eroded because he accepts the view that emotional disturbances arise from forces beyond his grasp. He can't hope to understand himself through his own efforts because his own notions are dismissed as shallow and insubstantial. By debasing the value of common sense, this subtle indoctrination inhibits him from using his own judgment in analyzing and solving his problems.[9]

It is indeed an attempt at indoctrination, though with generous intent and without the destructive implications that arise in efforts at political or religious brainwashing. The proper measure of it is not the impulse behind it but the effect. For me, visiting a dynamic psychotherapist was an encounter that made things very much worse, in two respects. Having an unshakeable conviction that I was evil and that my personality had broken on this realisation, I found the therapist's exasperation with me consistent with that premise. And her aversion to my looking elsewhere for psychological treatment deterred me from seeking remedies that might help. As I later found, these were available.

The search for release

Since my abortive psychotherapy sessions, I'd visited first a leading mental-health campaigner and then a doctor who'd recognised what I had and what needed to be done urgently. Unlike many other sufferers, I was fortunate to receive prompt medical attention, and unlike almost all I was able, through my work, to gain access to top-level professionals. The next person I called on was Dr John Williams, head of neuroscience and mental health at the Wellcome Trust. Having been put in touch with him through the Wellcome introduction, I went to his office for breakfast one day. He'd sent me the NICE guidelines on depression; I told him of my illness and my wish to know what science has found out about depression.

Dr Williams talked about research, drawn from many fields including genetics and imaging technology, into the workings of the brain – both how it normally functions and what happens when things go wrong. He explained that these new imaging techniques were building maps of how the networks of neurons are laid out in the brain. It was engrossing and I wanted to know more. My most urgent concern, though, was whether science could help me, now, understand why my brain wasn't working properly and what I could do about it. I spoke of my disastrous recent experience in talking to a therapist and asked whether there were better ways of dealing with depression.

He explained the advances made in understanding the brain, but also the limits of our knowledge. We don't know how the different parts of the brain interact with each other and create complex cognitive functions. But science makes strides in understanding mental disorder and in devising treatments for it. Referring to the clinical guidelines, he said there were psychological treatments that could aid recovery from mental disorder but that there was very little evidence that dynamic psychotherapy worked – except insofar as it is helpful, when in distress, to talk to someone who has a sympathetic personality. His advice reinforced the belief I'd already painfully arrived at: that counselling and dynamic psychotherapy may be useful for people who aren't ill, but are at best a diversion and sometimes (at least in my case) a danger to those who are.

For mental illness, something more is needed. There is no right answer for everyone, but there are plenty of wrong answers. Dr Williams advised me on what the science says, and that limited psychological treatments – the very things dismissed by psychodynamic therapists as tinkering with problems rather than solving them – are effective. He wished me success in getting out of the pit, with expert help, and suggested that whoever I saw should draw on different disciplines that had clinical trials behind them.

Stumbling upon effective treatment

I recalled that advice. Dr Nathan referred me to a clinical practice in the City but, by chance, I went through a directory of practitioners in central London myself and found one practice that provided mental-health treatment and psychological therapy. It was called The Mind Works and its list of specialisms sounded diverse. It seemed a more promising approach than the psychotherapist and I couldn't wait longer.

I booked an appointment for what turned out to be pretty much my darkest day of the whole experience. Arriving at the consulting room, I wept uncontrollably for no apparent reason. That took up much of the first session with a clinical psychologist, Dr Annemarie O'Connor. She kept saying it was OK and waited for me. She was younger than I'd supposed, in her thirties, and her youth relative to the psychotherapist was not as I'd imagined a guide would be. The 'natural priesthood' of middle age did not describe her. It was obvious to Dr O'Connor, as it had been to the psychotherapist, that I was in a state of acute distress. But there the methods diverged.

You don't need a doctor's referral to consult a psychologist, as (unlike psychiatrists) they're not classed as medical specialists. But a clinical psychologist will work very closely with a client's doctor, who is able to prescribe medicines.[10] Listening to me recount my experience with the original psychotherapist, Dr Nathan was certain it had caused damage and was anxious that I should see someone who could genuinely help. He pointed me in the right direction and made sure that he saw me weekly while I was having this more fruitful form of treatment.

Dr O'Connor's first question to me, when I'd subsided from the sobs, was whether I'd ever imagined going to a psychological practice for help. I explained about the psychotherapist and, not very coherently (in fact, more disjointedly than I'd managed with the therapist), all that had brought me to this state of helplessness. I recall little else of that first session except that, by the time we closed, I'd remembered to ask her – as Dr Williams had advised – what she did, in practice, and whether she used different approaches.

Dr O'Connor was sympathetic at my experience with dynamic psychotherapy. She saw little value in it, but said there was evidence and method behind other forms of talking therapy. These included CBT, compassion-focused therapy (CFT) and several others. I didn't really take it in, but I saw over weeks and months what she meant. Instead of searching my memories for a key, she showed me how to act directly on depressive thoughts and overcome them. Dr O'Connor literally taught me to think again.

Depression as cognitive error

This is how we proceeded. Every week I would describe not the minutiae of my past life but what I had thought and accomplished over the previous few days. The techniques Dr O'Connor taught me, principally but not only under the rubric of CBT, made no grandiose claims to uncover the role of the unconscious. They were more about getting me to talk to myself. That may sound undramatic, but it helped pacify me if I woke in the middle of the night and my demons came calling.

The treatment was not Freud, penetrating to the deep unconscious, but Socrates, considering and rejecting bad ideas.

In a process of dialogue on a mundane scale, I started to test and consider my destructive ways of thinking. Dr O'Connor explained that my depression was a severe illness, but not at root a mystery: it was born of cognitive error. Recovery, and guarding against a relapse, lay in my own hands. It required interrogating the beliefs that had caused my mental collapse and replacing them with better ones.

CBT is the treatment recommended under the clinical guidelines. It's an approach whose roots do in fact lie in the psychoanalytic tradition initiated by Freud – or rather, it's a response to that tradition's failures. Freud's theories dominated the US psychiatric profession from the early years of the last century (although his only visit to America was in 1909) to the 1960s. As often happens with intellectual movements established by a commanding personality, numerous splinters emerged soon after, each devoted to establishing the 'true' spirit of the master's intellectual legacy. The formal differences are arcane; the more obvious characteristic is what the various groupings had and have in common, which is a belief that mental disorder in its broadest sense is an affliction caused by the unconscious mind.

This theory had absolutely no empirical support. Indeed, there was no clinically validated treatment for mental illness till the early generation of antidepressants in the 1950s. Psychological treatments received no clinical support till the 1960s. That's where CBT broke new ground. It's grouped within the class of treatments known as psychotherapy because it, like them, is a two-way exchange between the therapist and the client. Moreover, while I had the benefit of a qualified clinical psychologist, CBT doesn't require a skilled practitioner. Even so, the differences between dynamic psychotherapy and CBT are substantial.

The development of CBT was in part a reaction to the failure of earlier approaches. Aaron Beck, the pioneer of CBT, had been trained in psychoanalysis and was initially a true believer. 'I have come to the conclusion,' wrote Beck to a colleague in 1958, 'that there is one conceptual system that is peculiarly suitable for the needs of the medical student and physician-to-be: Psychoanalysis.'[11]

Thoughts affect feelings

The essential insight of CBT is that our thoughts affect our feelings, and if our feelings are disordered, they can be progressively altered by shifting the ways we think. It's possible to build up good mental habits that can ease depression and eventually dispel it. As a young psychiatrist committed to the ideas of psychoanalysis, Beck hoped to put these on a rigorous scientific footing with a body of empirical evidence. Hence he devised an experiment to test the central idea of psychoanalysis about depression. By this means, he believed he could demonstrate that depression is due to an internalised hostility that is repressed and then directed against yourself. With other researchers, he compared the reported dreams of sufferers from depression with those of healthy people.

To Beck's surprise, the depressed patients reported less hostility in their dreams than the healthy respondents did.

Yet the image that these depressed patients had of themselves in their dreams corresponded quite closely with the way they thought of themselves when awake. Beck noted that, in interviews, his patients expressed exaggerated self-criticism and feelings of guilt. These negative thoughts were founded on catastrophising (that is, seeing an event or situation as dramatically worse than it actually is), and concluding from a single setback that this was how things would always be. The patients' judgements were strikingly divergent from the reality of their lives. Armed with these observations of depressed patients, Beck progressed from diagnosis to remedy. He encouraged his patients to examine the distortions in their thinking. Many responded rapidly to this treatment, to the extent that they ceased coming to him after a dozen sessions.[12]

The flawed thinking hypothesised by Beck described my state. My failings, failures and disappointments were voluminous. I knew that well enough. They accumulated in my mind and weighed upon me over years. They combined – they coagulated, while the mass expanded remorselessly – so that the setbacks became of dramatic significance. It was a type of solipsism in reverse. It assumed that of all the ills in the world, the most glaring and heinous were those that characterised my life and that I was responsible for them. Nothing was more corrupted and evil than my own nature, and I shrank from the realisation. My extreme loneliness had a ready explanation – one so obvious that I wondered why it hadn't occurred to me before – in the depravity of my soul. Everyone knew of this and therefore had no need to speak of it, to me or among themselves. Those close to me, the friends who knew me best, were fully aware of it even while they gave me refuge in their homes. So one week I stopped going to see them and didn't contact them, which caused them worry.

It's an example of how depression is distinct from conventional sadness even of an unusually extreme kind. The emotions of someone who has borne tragedy in life are intense to a degree unimaginable to those of us more fortunate, yet they are proportionate to their cause. Depression is a twisted kind of thinking. A cycle of isolation and of destructive thoughts feeds on itself and intensifies the severity of the depression. And it's a recognisable state. One of the greatest works of English letters, which explained the maladies of the mind according to the science of the day four centuries ago, has long been superseded in its analysis of their causes but never surpassed in its descriptions of the symptoms. In *The Anatomy of Melancholy*, Robert Burton recounted:[13]

> 'Tis my sole plague to be alone,
> I am beast, a monster grown,
> I will no light nor company,
> I find it now my misery.
> The scene is turn'd, my joys are gone,
> Fear, discontent and sorrows come.

The scene is turned. The thoughts that crowded into my mind were of my baseness. Emotions come from thoughts, and severe depression was their outcome for me. Beck developed

CBT on the principle that destructive thoughts can be identified and then modified. A sufferer from depression can be trained to anticipate when negative thoughts will arise – the situations, the times of day or night, the external stimuli – and consciously adapt their thinking. It doesn't require, or involve at all, a thorough search of all one's memories and recollection of life experiences. It focuses on the thoughts that are causing depression and on changing them. It breaks a cycle of self-destruction.

A scientific footing for psychological therapy

It works. And this is what distinguishes Beck's theories from what preceded them in psychotherapy. I don't mean only that his methods worked for me, whereas different treatments may work for other people and various types of personality. Rather, CBT put therapy on a scientific footing. Previously the field had been dominated by competing but essentially unresolvable claims about the therapeutic experience. Splinter groups of psychoanalysts sprang up, each with their intellectual inspiration and leaders, much like the recognisable sectarian tendencies of religious zealotry or radical politics. As Beck later described it:

> The mental health field is dominated by a few durable establishments and clusters of smaller sects of more tenuous standing. The major schools within this domain share certain characteristics: a conviction of the ultimate truth of their

> own system, disdain for opposing theories, and a steadfast emphasis on purity of doctrine and technique. In many instances, the popularity of a particular system seems to depend more on the charisma and single-mindedness of its originator than on the soundness of its foundations.[14]

This comment is understated and diplomatic, but Beck's judgement is an indictment of the psychoanalytic school from which he'd come. Against that approach, Beck insisted that therapy should rest not on assertion but on data and repeatable outcomes. That's what distinguishes the therapeutic method that he devised from the guesswork of the gurus of other schools of psychotherapy: evidence rather than reputation. The psychiatrist and neuroscientist Eric Kandel, a Nobel laureate in medicine, has summarised Beck's achievement this way: 'The crucial point is, Beck took a form of psychotherapy and he did a series of systematic, empirical studies that showed it's more effective than placebo, and that it's as effective as antidepressant drugs in mild to moderate depression. And he wrote a manual for the therapy, a cookbook, so that others could do studies as well.'[15]

By the time I had need of therapy, extensive studies showing the effectiveness of a cognitive-based approach to depression had been done. Beck had devised his own scale for measuring depression, known as the Beck Depression Inventory, and was a pioneer in quantitatively assessing the effectiveness of psychological therapies, the 'talking cures'. And so, after my false and potentially disastrous start in therapy, I was offered therapy that had been tested and validated.

This was the principal approach that Dr O'Connor followed in her sessions with me. Like the psychotherapist before her, she listened to my account of what had brought me low. Yet instead of delving further and further back into my history, seeking in it a clue to my present discontents, she began to introduce caveats and qualifications to what I'd said. She listened closely, but didn't just listen. This was a big difference from the psychotherapist, who'd been intent on seeing where my recounted thoughts took us. With CBT it was an exchange of thoughts and interpretations, according to a clear purpose. In listening to my immovable conviction that I'd failed every person close to me, Dr O'Connor introduced the notion that in any adult relationship there is a shared responsibility. It may not be equal, but there is never a monopoly of influence on one side. Whatever had happened, the reason for my loneliness was not a cankered personality, nor was its outcome an inevitable state of guilt and shame.

These were simple observations that began, very slightly, to affect how I thought. CBT can be done by anyone, or in a computerised form, but for me it was crucial to meet someone who I knew was amply qualified, with a doctorate in psychology, and whose personality I could relate to. I've readily granted that another form of therapist may be helpful if they provide a sympathetic hearing to a person experiencing problems in life. There's no science in what they practise, and for people with troubled personalities, and suffering major depressive disorder, they are not the right recourse. But the experience of communicating with a sympathetic personality, whom you can talk to freely and who is deeply interested in a client's experience, makes a difference. In therapy there is no substitute for dealing with someone you trust, not only in the sense of professional confidentiality but in ability and knowledge. If you don't feel trust, you can always go to someone else, and so you ought to.

The depressive catastrophises

The depressed mind will often engage in black-and-white, all-or-nothing rationalising. If my friends unusually omitted to check up on me, the explanation was clear: it was because they *knew*. They had discovered a truth that was impossible to hide. They were aware of my moral turpitude and had decided they could no longer trouble to disguise it. I explained this too to Dr O'Connor in the week of my greatest isolation. CBT doesn't involve the commonsensical reply of 'Don't be so silly', and she didn't give it. But she listened carefully and then responded. Her answer would instead, in this case as in everything else that exemplified my depressive thinking, test my belief against the available evidence and propose alternative explanations that had greater plausibility. Could the temporary silence from those close to me be due not to their knowledge of my wickedness but, perhaps, to the fact that I'd fallen silent myself, and my friends assumed I wished not to have my grief intruded upon? On balance, this did make sense. It was a hypothesis at least worth considering. The possibility that it was true subtly shifted my thinking not to a different view of the world but to a willingness not to assume the worst without limit.

That was how my sessions with Dr O'Connor proceeded. There was no deep scrutiny of my early life and my feelings about it, as a psychoanalyst might have engaged in. I don't recall ever referring to my childhood; it wasn't relevant, as I wasn't that person any longer. Instead, it was my beliefs *now* that were interrogated, to progressively test them and thereby, piecemeal, erode the depression that they elicited. The difference between popular and scientific approaches to mental disorder is encapsulated in this distinction. As one valuable dissection of the myths of popular psychology recounts: 'Extensive research demonstrates that understanding our emotional history, however deep and gratifying it may be, isn't necessary or sufficient for relieving psychological distress . . . In fact, treatments that place minimal emphasis on recovering or confronting unresolved feelings from childhood are typically equally effective as, or more effective than, past-oriented approaches.'[16]

My catastrophising, with one baneful belief built on top of another, had gone on for so long that it had become my default mode of thinking. I'd noted, for example, that I didn't have problems with performing my professional tasks, but that personal life was hard. Instead of trying to bring those parts of my life into balance, I concentrated on the gap between them, which became in my mind an unbridgeable chasm. In my familial duties and personal relationships I was constantly falling short and I couldn't work out why. I felt blazoned with failure, and this in my mind turned to shame and thence to ignominy. The explanation was, and had to be, that I was morally abhorrent and that the stain could never be lifted. Things would always be like this, and the weight of this knowledge would ensure that life could never again be a joy.

It was a mutually reinforcing concatenation of beliefs that ultimately triggered in me what is commonly known (though it isn't a medical term) as a mental breakdown and a severe depressive disorder. I'd assembled a set of facts, interpreted each to an extreme, and generated a conclusion to make sense of the whole. The conclusion was that my life till then had been a lie, that the trappings of success and satisfaction were chimerical, and that those I cared for and loved had suffered grievously from neglect born of my malignance. To compound my sense of shame and guilt, I'd hidden this discovery from everyone; yet I came to realise that they must already know it. This was how I reasoned in a depressive state. Every segment of the scheme is internally consistent and fits with the facts that I'd selected. If I'd recounted it to those close to me they'd have expressed incredulity, and this in turn would have convinced me I was right. This was how the world was, and I would never escape the infamy that I'd brought on myself.

The catastrophising was stark. Dr O'Connor listened to my facts and began to ask about their details. She explained how depression was an illness that arose from a particular way of thinking. And then we talked, week after week, about my beliefs and how we could interpret the same set of facts differently.

Healing the catastrophising mind

There would have been no purpose in simply contradicting those beliefs. We had to work through them sequentially, first by inquiring about the evidence I had for each and then suggesting alternative explanations for them. A disappointment in love, for example, may be evidence of irremediable character flaws that can never be repaired, but another possible explanation is that sometimes things which look right in the abstract just don't work in practice. The recipe, said Dr O'Connor, may specify all the ingredients in the right proportions but the cake may not turn out as you'd wish. Sometimes there is no explanation to hand, and the depressive mind is mistaken even in principle for trying to construct one. And even if one does exist, there is no useful purpose in seeking it. Likewise, my guilt at having not spoken as much as I felt I ought to have done to my father in his last illness need not be explained by failure, but rather by loss. He may have been quietly content at what I'd done in life, even if I was now certain that any such achievements were hollow.

I don't wish to depict this type of questioning approach and behavioural therapy as a remorseless stress on rationality. CBT, at least as I experienced it, is not this. The therapy isn't a system of answers. The therapist doesn't have any. And despite its methods of examining destructive ideas, CBT doesn't hold that you can reason your way out of depression. You can't; depression is resistant to logical thinking and doesn't work like that. Bad beliefs can be replaced with better ones but not by inquiry alone, precisely because depression is an *abnormal* emotional state. You need to *feel*, and not just give intellectual assent to, the truth of a counter-narrative in which there is no black and white but many shades of grey. It is as Stephen Crane, a pioneering American realist author of the late nineteenth century, described the voyage of life some sixty years after Thomas Cole had painted it:[17]

> When the prophet, a complacent fat man,
> Arrived at the mountain-top,
> He cried: 'Woe to my knowledge!
> I intended to see good white lands
> And bad black lands,
> But the scene is grey.'

CBT doesn't patronise the sufferer by telling them they're delusional. It's a method of testing the beliefs that underlie a depressive state. It's practical and immediate. It doesn't attempt to make unevidenced generalisations about the human condition and how it's driven by elements of the subconscious; it helps people who are ill, despairing and barely managing. It gently shifts the depressive mind towards considering that the scene is grey.

The compassionate mind

In a state of mental disorder, rationality is a hard goal to achieve. Sometimes there's no question of getting anywhere near it. Another technique the therapist will employ under these circumstances is CFT. This compassion-focused approach is associated with Paul Gilbert, a British clinical psychologist who has served on the committee establishing the NICE guidelines on depression. Its principles are set out in his book *The Compassionate Mind* (2010). Essentially, CFT promotes better mental health by encouraging sufferers from disorder not to reason their way to recovery, which cannot be done, but to be compassionate towards themselves and others. CFT can admittedly have a weird quasi-spiritual terminology that makes me wary. It speaks of healing and it draws on Buddhist philosophy. When I read up on the subject, I was initially suspicious too that CFT looks into a patient's distant past in order to explain their current mental disarray. But it is, I found, a valuable adjunct to other behavioural therapies.

CFT theory suggests that humans have evolved basic emotional responses in order to survive. These are awareness of threats and an instinct for self-protection; a drive to attain goals and resources; and the contentment of being part of a community. Gilbert argues that a state of mental calm is attained by an awareness of being socially connected and safe. A sense of threat may come from childhood experiences of insecurity – of having been failed by the 'attachment figure', a parent or guardian that we trust. Thus, according to a review of the literature of attachment theory, 'safeness [in Gilbert's scheme] itself is not just the absence of threat, but the product of a specialized brain system, tied to social cues, that is regulated by different neural pathways than those involved in responding to threats'.[18]

What does this mean, in practical terms? It suggests that people can have their sense of threat activated merely by lacking reassurance. That sense of foreboding can conversely be deactivated by the exercise of compassion, by others and by the sufferers themselves. And this is an essential insight for those in the grip of clinical depression, because they are afflicted with overpowering guilt and shame.

What appealed to me about CFT, despite its terminology, was that it had theory behind it. I could understand what it aimed for and how it worked. This is different from the Buddhist ideas that in part inspire it. In the corporate world of the twenty-first century it's hard not to come across the allied notions of mindfulness and meditation, and the claimed benefits of them for mental and physical wellbeing. Meditation has been practised for millennia, but modern neuroscience has found only limited supporting evidence for its proclaimed effects in easing stress. A meta-review published in 2014 by the US Agency for Healthcare Research and Quality found 'either low SOE [strength of evidence] of no effect or insufficient SOE of an effect of meditation programs on positive mood, attention, substance use, eating, sleep, and weight. In our comparative effectiveness analyses, we did not find any evidence to suggest that these meditation programs were superior to any specific therapies they were compared with.'[19]

Research continues, with an expansion of the number of randomised clinical trials involving mindfulness meditation since the meta-review was written. The resulting evidence was summarised in 2018 by Gaëlle Desbordes, a neuroscientist with the Depression Clinical and Research Program at Massachusetts General Hospital: 'There are a few applications where the evidence is believable. But the effects are by no means earth-shattering. We're talking about moderate effect size, on par with other treatments, not better. And then there's a bunch of other things under study with preliminary evidence that is encouraging but by no means conclusive. I think that's where it's at. I'm not sure that is exactly how the public understands it at this point.'[20]

Despite its claimed benefits, mindfulness meditation has only limited empirical evidence for its effectiveness and no clear explanation for how it can promote mental health. Compassion rather than mindfulness makes sense. Depression is a state of debilitating isolation and loneliness. To receive compassion from others may not be enough (and wasn't in my case) to dispel it. Compassion alone wouldn't have tackled the belief system that had left me stricken with horror at what I had become. But compassion towards the self did, in my experience of coping with depression, have a valuable buttressing role. By definition, the powers of rationality of the depressive mind are attenuated. Recovery can't depend on them. So in addition to our mutually examining the mental presumptions I held, Dr O'Connor advised me to imagine things that would soothe my fevered mind. It could be something as simple as a hug from a loved and trusted friend, or the embracing waters of a swim on a balmy day. As we assessed each week my depressive state and any changes I felt in it, she pressed me to focus on things I'd achieved even while mentally debilitated.

This was practically how I learnt of psychological treatments for depression. I experienced what the clinical literature said. Methods seeking incremental improvement and based on evidence were valuable for me. Those that aspired to find the key to my long-standing unhappiness ended up intensifying the burden of guilt on the oppressed mind; for me, they were not only futile but dangerous. By abandoning the latter, I was better able to live with depression and emerge from it.

8

LIVING WITH DEPRESSION

> Friendship, peculiar boon of Heaven,
> The noble mind's delight and pride
>
> Samuel Johnson, 'Friendship: An Ode', 1743

I benefited from a dawning understanding that, through the blackness, I wasn't alone. Friends and colleagues, and sometimes even complete strangers, couldn't help but discreetly convey the solidarity of pity while solicitously carrying on as if I was normal. However well they knew me, they were diffident to broach the subject and I told only a handful of them what was going on in my head.

The oddities of my behaviour, with its alternate agitation and listlessness, were inescapable when I was in company, though I could for a while still give a reasonably convincing mimicry of normality when I was working and performing a role by writing an article, giving a speech or doing an interview. It was not possible to hide from those who knew me the disturbance of my mind and the haggard deterioration of my appearance. I looked distracted and stared into the distance. I said nothing unless spoken to, and sometimes not then either. When seated, I assumed an instinctively foetal posture to make myself dwindle; when standing, I stooped; when walking or engaged in any movement, I was sluggish and unsure. My friends couldn't explain and didn't know what was wrong, and they noted with hurt that I was almost always heedless of their concern and collective efforts to extract me from the mire. But they weren't heedless of me.

Nor, I found, was I solitary or at all exceptional in experiencing depression. The paradox of depressive disorder is that it's widespread far beyond what most people realise (and certainly what I was aware of), yet it's scarcely visible in the public domain. In talking and writing about depression since my experience of it, I've found it commonplace for people to know of the condition by having observed it in a relation, friend or loved one. Yet sufferers do not like to own up to mental illness even if they are aware that their moods are the outcome of it. In seeking to find a way out of depression, it is as well to remember that others are willing it too. And there are strategies and techniques for returning to them, restored to good mental health.

Shifting bad ideas

I've described in the previous chapter the aims of evidence-based psychological therapy in correcting distorted thinking. Yet it's crucial that, in examining the burden of destructive ideas and overpowering feelings of failure, the mental adjustment be credible. There's no purpose in trying to supplant clinical depression with feelings of joy and fulfilment. It won't work because the ideas won't be real. Like the temporary high of a couple of glasses of wine, they'll wear off, and the disillusionment, shorn of that fleeting and artificial emotional support, will be punishing. Compassion allowed me to live with myself even while I recoiled from the knowledge of my baseness.

As months passed, and I continued to see Dr O'Connor weekly, we devised strategies for me to cope with day-to-day tasks. The most basic of activities, such as opening the front door either to go out or to return home, were hard and I'd have to build up to them. Just to leave home in the morning was an ordeal, so I'd separate it into stages.

Indeed, everything had its stages, from the moment I awoke. As the sense of despair enveloped me, I knew that it would simply worsen if I stayed where I was and pulled the covers over me. To get out of bed required shifting my feet on to the floor and leaving them there till I was ready for the next stage. It was a desperately slow procedure. I'd tell myself that I had to make something of the day. It was by the instinct of living, as Spinoza had surmised was our state, that I proceeded. Any break in routine would have thrown all into confusion. It would have disturbed the habits by which I operated. Having showered, shaved and dressed, I'd always have coffee, toast and marmite, though my breakfast could have been anything: whatever I ate was invariably as ashes in my mouth.

To leave home and get to work, I'd sit on the stairs and wait till I was ready to open the front door. This could take an hour or more. When I got to the office I would switch on the computer and count it an achievement. Sometimes I couldn't write anything and concentrated hard on switching the computer off again. Often that was an entire day's work. All I could do, on the professional advice I had, was to treat it as normal – not only normal but a success – and try again the next day. My colleagues connived in the fiction.

Progress was excruciatingly slow and I had many setbacks. The toughest of these was when I would travel for a weekend to visit my children. I'd set off hours in advance to make a flight, knowing that I would need to stop many times on the way just to sit and practise the techniques I'd learnt, of mental calming and questioning. When I saw the children, I would try not to alarm them with what I knew was a demeanour they'd never seen before. I became practised at taking them on excursions where they would be safe on their own for a short time, like the dolphin show at the zoo, while I would wait outside and subside into the park bench with my head in my hands. It became too much by the end of the weekend, when I would sit on my own in the airport terminal and feel sick with misery.

I tried, as far as I could, to keep professional engagements and operate again by instinct, the observation of Spinoza that this is our state, rather than calculation. The strategy wasn't always successful. I'd agreed, many months before the illness struck, to take part in an Intelligence Squared debate on the 'language wars', alongside the classicist Mary Beard and against the journalists Simon Heffer and John Humphrys.

I went through with it. My case was that the English language is in rude health and that the complaints of sticklers and pedants about a supposed decline in linguistic standards are groundless. As I sat on the stage and listened to the first speaker (John Humphrys), I realised I'd forgotten any point I'd intended to make and didn't know what to say. I looked out across the auditorium in desolation. I never speak from notes. When it came to my turn to speak, instinct had to suffice.

In private, too, my faculties of memory and reason could depart abruptly. I took a day off work to make final revisions to a book on language, whose text I'd managed to deliver to my publisher just before the deluge of depression. With the proofs now in front of me, I could recall almost nothing of the book or its argument. By evening I found I'd frittered the day away by looking at suicide websites, not with any view to self-harm but because this was all I could think about. I was more than usually burdened by the sense that I had betrayed and failed those who depended on me, and that my moral culpability was known to all. The conviction ate me up from the inside and hollowed me out. It did strange things to my mental state. I had the notion that if only I knew in principle there was a reliable and painless way out of this vale of tears, then the knowledge, even though never acted upon, would calm me. It didn't (and there isn't). The day ended with one of those bizarre and scarcely coherent emails to friends who cared for me but couldn't reach me.

Extending the strategy for recovery

As part of the recovery plan devised with Dr O'Connor, I'd email a different friend each week whom I hadn't seen for a while, to get back in touch and say they'd been on my mind. If they were in London we'd arrange to meet – and often, retreating in anxiety, I'd cancel at the last minute. It must have seemed, and indeed was, very discourteous. One of them had the presence of mind to gently insist I turn up for our dinner reservation anyway. She thereby did me an immense kindness. I told her all that had gone wrong and realised that, like many others, she understood what this illness was because she'd seen people who were clinically depressed, though in her home culture in Sri Lanka they were often reluctant to speak of it. The commonness of the scourge was a revelation to me.

The same pragmatic approach to recovery governed my tendency to relapse into those earlier surmises about the permanence of depression and my ineradicable wickedness. You can't think your way out of depression: this was the message impressed upon me. You can follow every step in a train of thought contradicting you, but still be immovable and unconvinced. CBT is a way of testing bad beliefs and replacing them with better ones, but it's a process rather than a revelation. You have to first feel your way out of the pit in order to emerge into the daylight. If you try to reason your way out, you're likely to get stuck in rumination – an iterative process of all the destructive notions that caused the breakdown in the first place. It's not helpful to do this.

And because 'it's not helpful' is a constructive belief, Dr O'Connor showed me methods to apply it. If I woke in the middle of the night paralysed with anguish about why I was in this state and what had caused it, the right response was not to try and work it out by mental investigation. That was a hard lesson to absorb, because I desperately wanted to know why my mind had collapsed, and thinking hard about this conundrum might have yielded at least a clue. But the more constructive approach was to tell myself that, even if that clue was available through the power of sustained thought and recollection, I wouldn't find it, let alone reach an answer, here and right now. Perhaps I did need to know and one day would work it out, but that was for then. It wasn't for now. As my mind and body flailed, so the message I needed to imbibe was that I should postpone any such reckoning. Perhaps, I told myself, I'll think of it again tomorrow, or at the weekend, or the week after. There's time to find out.

By pushing back the deadline for my investigation, the salience of the questions that troubled me receded. It was an ordered, effective response to dealing with irrationality born of severe illness. I still do the same thing, long since I was cured, if an intractable issue troubles me and when I know it has the capacity to disrupt what I'm doing. There is time, I tell myself, and there is no immediate answer to be found, so it's best to postpone the inquiry and do something else – something more capable of resolution and therefore more helpful to everyone.

It changed the habits of my thinking. I came to understand that the causes of mental turmoil are not the events that trouble us but the weight we place on them. Take something, anything, that is a part of life yet hard to cope with: the experience of failed relationships, marital break-up or plain rejection. It's always distressing, at the very least because of hopes that were even involuntarily invested and that will never now be realised. You'd expect a stressful rupture to cause sadness and emotional turmoil, and any compassionate person would wish to help a friend or relation whose heart has been broken. But we can still stand apart and make a judgement on whether the relationship was suitable and stable. Sometimes it won't have been, and we'll adopt a pragmatic though unexpressed view that it may have ended for the best. Even if the partnership was on the face of it a good match, we can reasonably expect that the intensity of the pain will lessen over time. What causes sadness to persist, beyond an inevitable period of shock and grief, in someone who's been deserted by their partner or spurned by their loved one is the interpretation they place on the break-up. If their life's ambitions have rested on growing old with the one they love, then the emotional disruption will be extreme. It may stimulate a series of negative thoughts about their culpability and failings: *I was negligent*; *I failed to see the signs of a break-up*; *I don't understand why this happened, when everything seemed perfect and we have so much in common*; *I can never be happy again.*

The need to query private interpretations

That is where conventional sadness, albeit of an intense kind, becomes disorder. It's not the event but the value accorded to it and the interpretation made of it that cause this depressive cycle. As Beck has written of these private meanings, encountered in sessions with his patients (emphasis added):

> At times we find that a person's reactions to an event are completely inappropriate or so excessive as to seem abnormal. When we question him, we often find that he has misinterpreted the situation. His misinterpretation comprises a web of incorrect meanings he has attached to the situation. Interpretations that consistently depart from reality (and are not simply based on incorrect information) can be justifiably labeled as deviant . . . [T]he deviant meanings constitute *the cognitive distortions that form the core of emotional disorders.*[1]

Faced with a shocking and unanticipated event, a thoughtful person will naturally wonder what caused it and seek answers. Yet the answer that these interrogations will yield may be, as Beck puts it, deviant. It will be so if it departs from reality and makes sense only in a private interpretation. The role of the therapist here is crucial, and was so to me. Again, as Beck puts it: 'Private meanings are often unrealistic because the person does not have the opportunity to check their authenticity. In fact, when patients reveal such meanings to their psychotherapist, this is frequently the first chance they have had to examine these hidden meanings and to test their validity.'[2]

The therapist (in my case and to my good fortune, a trained clinical psychologist) assists in examining those meanings. The discipline proved invaluable to me in helping overcome a pattern of remorselessly logical but destructive and irresolvable thought. I had the experience of trying to address and understand disruptive events in my own life. Sadness over loss is natural; rumination over the causes of extreme sadness is a temptation but rarely one that's worth pursuing. And that's what I've learnt and come to accept. There are many disappointments in my life and flaws in my character, along with ambitions that remain to me, but I no longer fret about failures, real or imagined.

Depression and the solitary burden

Among those facets of life that I could wish I'd attained is the ability merely to share it. In the maelstrom of modern society there is an intuitively appealing quality in being solitary. That's just as well, as the number of single-person households is rising. Those in the UK living alone exceeded eight million for the first time in 2018. Around 15 per cent of the adult population now live on their own, and there's been a particular increase in people like me – those in middle age who are single. Partly that's because there's been an increase in the proportion of the population aged forty-five to sixty-four, but there are other factors too. Male life expectancy has been catching up with female life expectancy, and there's been a rise in the proportion of the population who are divorced or who have never married.[3]

Yet the romantic ideal of fused sensibilities and emotions was what I would have wished. Like the hapless amateur in any field, I'd read books about it.[4] A real partnership, a meeting of hearts and minds, is reciprocal. Love is more than a fleeting *coup de foudre*, though an unintended yielding of the emotions is invariably how it starts. To be real, and not merely sustained, it has to be a rational, well-grounded recognition of real qualities in the person one loves. The ideal is expressed in Dante's love for Beatrice, recounted in the final section of the *Divine Comedy*, as he travels through Paradise:[5]

> Again, mine eyes were fix'd on Beatrice;
> And with mine eyes, my soul that in her looks
> Found all contentment.

There was nothing more that I wanted from the emotional life. I have never found it. Solitude, where the pursuits of the life of the mind combined with the succour of single parenthood, provided its own rewards. The difficulty comes when solitary introspection turns to frustrated rumination and, in the absence of an answer, then to untrammelled horror at what one finds within. This was what happened to me. Exhaustive and exhausting rumination brought me to that state. My way out was to stop continually seeking answers to conundrums that had none, to learn that my experiences were far from exceptional, and then slowly alter the catastrophising way I interpreted them. I learnt that the question I posed to myself was essentially unanswerable, and that the most helpful approach was to cease trying to address it.

To inch back the darkness

Slowly but discernibly the darkness around me began to dissipate. At first it was sporadic and very occasional. One evening I got the bus home from the City and, as it passed the Palladian elegance of Shoreditch Church, I quite suddenly had a sense not of peace but of recollection. I could recall what it felt like not to have depression. It was fleeting, lasting maybe a few minutes, but it was a profound experience. I'd found that the darkness was not absolute; depression was not a monolith. It had its weaknesses. It could be eroded. The scene was grey rather than pitch. Perhaps if I waited long enough it would lift entirely, or at least to an extent that I'd be able to live with and survive. Maybe, in infinitesimal stages that I wasn't aware of, it was dispersing already. It felt as if someone had reached out a hand through the gloom and clasped mine before withdrawing. A religious person, unlike me, might have seen in it an act of grace.

It was a minor episode, but a turning point. After a prolonged and painful effort, I began to take pleasure in small things once more. Dr O'Connor had advised me to resume reading books. If I got only ten per cent of my previous enjoyment of a Wodehouse novel, or I took in only a few words on the page, it was an advance. Reading books whose contents I already knew, through the familiarity of countless readings, invited a sense of comfort – of being at home in the world, with one or two fixed points of reference that I could share with my past. I did as she'd suggested, but it was hard. The words swam on the page. I could recognise them individually and in phrases but still make little sense of the syntax or semantics. Yet as I persisted, the idea of recreation returned. I could maintain concentration for a little longer each time.

The same was true of music. I had given up listening to anything, as it seemed cacophonous and made little sense. I developed an intense aversion to Mahler, which has remained, not for any particular reason or bad experience, but because of the ambitious extent of his symphonies. I felt dwarfed, intimidated and repelled by their scale and the composer's wish that a symphony be like 'the world: it must contain everything'. I didn't want to be in the world or to examine it. I yearned to escape it. Yet a day came when I was invited by friends to a concert, Steven Isserlis playing Elgar's Cello Concerto at the Royal Festival Hall, and I didn't pull out.

In a pre-depressive state I'd known the piece well and always relished the sound of the cello. Like many German Jews in the nineteenth century, my family came off the boat at Liverpool and settled there. Max Bruch, the German Romantic composer whose First Violin Concerto has become a staple of the repertoire, served several seasons at the time as Principal Conductor of the Liverpool Philharmonic Orchestra. His *Kol Nidrei* for cello and orchestra, which he completed in Liverpool in 1880, is one of his few other works to be well known and it became a work of significance for my family. Though Bruch was not Jewish, he composed it explicitly as an adagio on Hebrew melodies. I had a great-uncle who told me, regarding this piece, that he considered the cello, with its mellow timbre, to be the orchestral instrument most akin to the human voice. Listening to the Elgar concerto, I found the same.

It was simultaneously like hearing music for the first time and returning to the social world of connection and conviviality. Again, it was a moment not of recovery but of realisation. I wasn't alone. I had to carry the burden of my guilt, but being among others could make the journey lighter.

9

DEPRESSION AND ART

> In isolation, man remains in contact with the world as the human artifice; only when the most elementary form of human creativity, which is the capacity to add something of one's own to the common world, is destroyed, isolation becomes altogether unbearable . . . Isolation then becomes loneliness.
>
> Hannah Arendt, *The Origins of Totalitarianism*, 1951

An avenue on the world

To be able to read once more, and listen to music, was a pathway to my gradually returning to a more normal state, and then a 'new normal' state. It was also a means of finding out about the illness I had been experiencing. I wanted to be able to describe it and, through the reach of the written media, to make some mark on public understanding of depression. Fortuitously I had an example before me that I could visit regularly.

Off Fleet Street, from which the newspaper industry has long since decamped to be replaced by law firms and banks, is a quiet square with an eighteenth-century townhouse in one corner guarded perpetually by a cat. The house was Samuel Johnson's; the cat is a statue of his feline companion, Hodge. Inside are sparse but elegant furnishings from the great lexicographer's estate, and a facsimile of his *Dictionary of the English Language* of 1755. I knew his work well and had very recently, in writing a book about language, read up too about the significance of his dictionary to English letters. I'd often visited his house in London and occasionally his birthplace, now a museum, in Lichfield. If I could get to the house in lunchtimes when I was on my own, I did. It calmed me in my labours and unhappiness. The evidence of Johnson's industry and of his leisure made me feel at home, but so now did his personality. He was tortured by a sense of guilt and he desperately sought expiation.

Johnson's work on words was a scientific enterprise. This isn't usually the way language is seen. In my profession, pundits typically acquire the status of style 'gurus' merely by force of assertion rather than specialist knowledge. Scholars of language do it differently: they will only arrive at conclusions on the grammar of a language and the semantics of its vocabulary after looking at the evidence of usage. But whereas knowledge of the external world is derived from something real and material, without and within, words are mental constructs. Their only existence is through a social contract under which combinations of sounds correspond to thoughts, images and ideas. Words have no intrinsic connection with their meanings, yet those meanings are common to every speaker of a language. Johnson set about recording the way that language is used. His dictionary was a revolutionary undertaking because it exemplified the principle that words' meanings are what we, the users of language, make of them.

In the mind, yet also in the world: such is language. Johnson's industry codifies it not by edict but by discovery. He read extensively and discovered what words meant from the way they were used by Shakespeare, Bacon, Milton, Newton and many others, whom he quoted liberally. Language makes us at home in the world by allowing us to replicate discoveries and convey ideas across societies and epochs. A dictionary brought me home by exemplifying that faculty. And the mental struggles recounted by James Boswell in his *Life of Johnson*, the first modern narrative of personhood in all its fascinating idiosyncrasies, finally made complete sense to me too. I dipped into it anywhere and read late in the evenings, when I took my drugs.

Johnson's depression

Depression, the 'morbid melancholy', was Johnson's affliction throughout his life. As Boswell told it: 'While he was at Lichfield, in the [Oxford] college vacation of the year 1729, he felt himself overwhelmed with an horrible hypochondria, with perpetual irritation, fretfulness, and impatience; and with a dejection, gloom and despair, which made existence misery. From this dismal malady he never afterwards was perfectly relieved; and all his labours, and all his enjoyments, were but temporary interruptions of its baleful influence.'[1]

Johnson's sense of constant guilt for falling short was the mark of a fellow sufferer from depression, on a grand and terrible scale. The account appalled and resonated. 'This is my history; like all other histories, a narrative of misery,' he wrote to his friend Bennet Langton at the end of his life.[2] The misery was a roster of physical ailments of an ungainly body worn out by exertion. Yet the abiding torment and constant companion was something else and more severe. The condition was, as Boswell recounted it, 'that constitutional melancholy which was ever lurking about him, ready to trouble his quiet'.[3]

The disorder never left Johnson and was observable in its effects. One of his old friends recounted finding him 'in a deplorable state, sighing, groaning, talking to himself, and restlessly walking from room to room. He then used this emphatical expression of the misery which he felt: "I would consent to have a limb amputated to recover my spirits."'[4]

It was the same: the disorder and the turbulent spirits, which with me were translated into lassitude rather than pacing, and frozen horror more than voluble lament. The cause was common too: an overwhelming, crushing ineradicable guilt born of failure. The sources of Johnson's anguish were numerous, but they appear to have derived from his fear that he was squandering his life through idleness and frivolity. Even after completing the dictionary, a labour of nine years, he wrote a poem in Latin, which in translation by

Arthur Murphy is titled 'Know Yourself', in which he faced the future with dismal apprehension:

> The listless will succeeds, that worst disease,
> The rack of indolence, the sluggish ease.
> Care grows on care, and o'er my aching brain
> Black melancholy pours her morbid train.[5]

It was guilt for what he took to be his indolence, heedlessness of those close to him (especially after the death of his wife), and sins of impurity that the pious mind is ever susceptible to. I could recognise what it meant now to be crushed by the guilt and shame arising from the immovable conviction of failure. But what I knew of Johnson's source of solace was neither effective for him nor open to me: the comforts of religion. For Johnson, 'Real alleviation of the loss of friends, and rational tranquillity in the prospect of our own dissolution, can be received only by the promises of Him in whose hands are life and death, and from the assurance of another and better state, in which all tears will be wiped from the eyes, and the whole soul shall be filled with joy. Philosophy may infuse stubbornness, but Religion only can give Patience.'[6]

Johnson's sufferings were not, after all, alleviated this way, through religious faith. On the contrary, he was stricken with terror at the apprehension of death and was incredulous at the notion that the rationalist David Hume had succumbed to it without torment and with equanimity. It troubled him deeply. Such was the account of Boswell, however, who, according to his narrative of his last meeting with Hume, 'asked him if the thought of annihilation never gave him any uneasiness. He said not the least; no more than the thought that he had not been, as Lucretius observes.'[7]

Emancipation from religion and submission to it had radically differing effects on the respective mental states of these two great figures of eighteenth-century letters. It does not follow from Hume's equanimity that free thought is the route to mental health, for depression and disorder can strike anyone. In a challenging essay on the relation between Hume and Boswell, and their respective attitudes to religious consolation, Michael Ignatieff argues that Hume was a case apart in not needing it, for he was at peace with himself. Conversely, Boswell (even more than Johnson) could barely stand himself. 'We have needs of the spirit,' says Ignatieff, 'because we are the only species whose fate is not simply a mute fact of our existence but a problem whose meaning we attempt to understand.'[8]

There is truth in the metaphor (for that's what it has to be) of 'spirit'. Beyond doubt, the Stoic virtues of self-mastery that Hume displayed, and which shocked Boswell, proved impossible for me to attain. I had aimed for reason and equanimity in a purposeless universe, yet acquired instead the same sense of self-disgust that dogged Boswell throughout his life. It's not fanciful to suppose that Boswell sought deliverance from this state through his attachment to two of the wisest figures of the age, and indeed of any age, in Johnson and Hume.

Yet Johnson's ceaseless suffering does at a minimum demonstrate that deep religious faith can coexist with severe and lifelong depression. Submission to God is not a route to happiness. Johnson went so far as to dispute that happiness in the present was even possible. As Boswell tells it:

> He asserted that the present was never a happy state to any human being; but that, as every part of life, of which we are conscious, was at some point of time a period yet to come, in which felicity was expected, there was some happiness produced by hope. Being pressed upon this subject, and asked if he really was of [the] opinion, that though, in general, happiness was very rare in human life, a man was not sometimes happy in the moment that was present, he answered, 'Never, but when he is drunk.'[9]

While enduring agonies in the present, Johnson held out the possibility of hope for the future. But this doesn't suggest that faith will alleviate misery. The dark night of the soul, in which the pious struggle with doubt, is easily mutable into the darkness of depression. Johnson's life was a succession of sufferings, of which the sense of guilt and failure was most acute. His descriptions are so evocative that I recognised readily the sense of failure and futility that summarised my life, the realisation of which had come upon me suddenly and unexpectedly. As Boswell recounts his friend's observation: 'When I survey my past life, I discover nothing but a barren waste of time, with some disorders of body and disturbances of the mind, very near to madness, which I hope He that made me will suffer to extenuate many faults, and excuse many deficiencies.'[10]

Johnson's experience exemplifies what I observed about depression, in myself and in the accounts of many others. It is a common human suffering, and we tend to interpret and reinterpret it in the language most familiar to us. The triggers for it are as complex, various and heterogeneous as the sufferers themselves, but the uncaused and essentially irrational nature of the ailment is a constant. Depression involves a loneliness taken to such an extreme that the normal bonds of human understanding and fellowship are severed.

Burton's melancholy

Johnson found reward – perhaps enlightenment, but more probably simply the comfort of recognition – in Robert Burton's *Anatomy of Melancholy*. In the recollection of an Irish cleric, the Rev. Dr Maxwell, a long-standing friend of Johnson: 'Burton's *Anatomy of Melancholy*, he said, was the only book that ever took him out of bed two hours sooner than he wished to rise.'[11]

It is the greatest of Renaissance accounts of depression, and the science of Burton's *Anatomy* reflected the understandings of the times. As Andrew Scull summarises it, in an extensive cultural history of insanity: 'Like his medical forebears (from whom he quoted extensively), Burton viewed melancholia as generally the product of an imbalance of the humours, and especially a superfluity of black bile.'[12]

The four bodily humours, or fluids, were a concept inherited from the ancient Greeks, principally Aristotle,

Hippocrates and the physician Galen of Pergamon. These were thought to be physical qualities that determined the behaviour of the body and indeed of all creation, reflecting the four elements of earth, air, fire and water. The humours comprised black bile, yellow bile, phlegm and blood, which corresponded to the four temperaments of the melancholic, the choleric, the phlegmatic and the sanguine. An imbalance of the humours caused these temperaments to become pronounced, of which the melancholic was the least desirable yet also the most creative. The medical solutions to a supposed imbalance included the common practices of lancing and draining blood through the use of leeches. It was grim, such that, according to Burton, the people complained: 'That the State was like a sicke body which had lately taken Physicke, whose humours are not yet well settled, and weak[e]ned by so much purging, that nothing was left but melancholy.'[13]

What we gain from reading Burton is not the medical diagnosis but the sense of what melancholy is, and the poetry in which the condition is couched. He expresses the melancholic temperament of the lone traveller in life: 'Voluntary solitarinesse is that which is familiar with Melancholy, and gently brings on like a Siren, a shooing-horn, or some Sphinx to this irrevocable gulfe.'[14]

The myth of melancholic creativity

Burton hinted at a creative role for melancholy, declaring that those with the condition are the most 'witty' (that is, intelligent and with their wits at the ready). He speaks of 'that drunkennesse which Ficinus speakes of, when the soule is elevated and ravished with a divine taste of that heavenly Nectar'.[15] The scholar he cites, with his Latin name, is usually known as Marsilio Ficino, a Roman Catholic cleric and philosopher of the early Italian Renaissance in the fifteenth century, who did much to spread the notion that a state of melancholy was a spur to learning.[16]

Ficino's reasoning came from a discipline that is entirely fanciful, namely astrology and the supposed influence of planetary alignments on human character. And the conclusion he drew from it is alarming. The modern sufferer from depression could not count it a heavenly nectar. Not at all. And this is another pragmatic reason why the division, urged by some psychologists, between depression and 'melancholy' is not likely to take hold in the public mind.

As we've seen, the argument is that different terms are needed for clinical depression, which is an abnormal emotional state, and intense sadness, which is what everyone undergoes at some time of stress or loss. The term 'melancholy' won't serve for the latter, though, because of the weight of its connotations. In literature, it denotes a pleasing state of introspection rather than extreme sadness, let alone ravaging illness. To be cast down with melancholy, according to Ficino and other Renaissance thinkers, was to enter into a greater communion with departed generations and be more attuned to the workings of divine providence. This notion was itself derived from antiquity, specifically Aristotle's conception of the reflective genius: melancholy was the outpouring of the love of learning. It was the opposite of the humanism promoted by the Age of Reason, and it thrived under the illusion that poetic genius depended on tragedy and solitude.

You find this ideal of melancholy in Shakespearean comedy as well as the tragedy of Hamlet. In *As You Like It*, Rosalind says to Jaques: 'They say you are a melancholy fellow.' He replies: 'I am so; I do love it better than laughing.' But in the context of the play it's intended as a joke, and Jaques, as the butt of it, explains that his emotional state is a 'melancholy of mine own, compounded of many simples, extracted from many objects, and indeed the sundry contemplation of my travels, in which my often rumination wraps me in a most humorous sadness'.[17]

This parody in Elizabethan drama was well founded. It rebelled against the enduring notion that melancholy was akin to pensiveness, and that it heightened a writer's sensitivity to the mysteries of existence. Shakespeare's mockery sadly wasn't effective, judging by the work of later literary figures. The very term 'melancholy' continued to epitomise the ideal of the poet who, by heroically delving into ultimate questions, escaped the trivial constraints of normal life. John Milton wrote in 'Il Penseroso' (1631), his lyric poem on the theme of melancholy as a stimulus for contemplation and art:[18]

> But hail thou goddess, sage and holy,
> Hail divinest Melancholy,
> Whose saintly visage is too bright
> To hit the sense of human sight . . .

What was satire in Shakespeare became a real intellectual force of the Augustan and then the Romantic age. The love of melancholy became an entire cultural movement of Gothic poets reflecting with delighted self-absorption on the miseries of the human condition.

The greatest of these later writers was John Keats, in his 'Ode on Melancholy'. The poem is not autobiographical; it is a reflection on the state of how, 'in the very temple of Delight, / Veiled Melancholy hath her sovereign shrine'. His point is that only those who have known the melancholic disposition can truly savour joy and also understand the loss of it: 'His soul shall taste the sadness of her might, / And be among her cloudy trophies hung.'[19]

Keats's characterisation of melancholy is lucidly explained by the literary scholar Robert Cummings. The crucial point is that pleasure and pain are necessarily bound up with each other: 'The extremity of delight is generated by a suffering at the center; and the profoundest exploration of delight returns us to its origin in grief. It is a fact of human experience that desire is painful, and our arrival at the limits of pleasure includes that pain, not as following on pleasure, but because pleasure – even in its further reaches, or especially there – still includes desire.'[20]

Again, this is a conception of melancholy that is not only distinct from clinical depression, of the extremity that Johnson suffered, but that ties melancholy to the experience of delight. It's psychologically as well as poetically profound, but Keats is not writing of disorder: he's writing of what, in Unamuno's phrase, is the tragic sense of life. Where

Burton in the *Anatomy* hints at the pleasure of melancholy, he's also realistic about the agonies of what we now know as depression. He makes clear that he's suffered them. As Scull summarises him: 'Perhaps Burton's own melancholic temperament encouraged him to laud melancholia's connections to creativity, though he was certainly intimately familiar with the paralysing depression the black humour could bring. As he affirms, "that which others hear or reade of, I felt and practised my selfe", and while "they get their knowledge by books", he commented wryly, "I mine by melancholizing".'[21]

I'd learnt this much during my own mental paralysis and my inquiries into it. The modern understanding of depression is grounded in Burton's observation of its destructiveness. It's good and compassionate that we've moved on. The depiction of depression in letters may not be accurate in the scientific understanding, but it's historically vital. Because the workings of the mind are mediated to us rather than conveyed directly, we understand mental disorder only as it's described by its sufferer. Depression is real, and not imagined; of this we can be sure, as the experience has so many common facets across ages and eras. Its origins are physical, because everything – including cognition and thoughts, good and bad – is the product of material forces.

The malady of being: the case of Emily Dickinson

Yet the realm of art and letters is crucial to understanding depression. I found it expressed most acutely in a poet I'd read for many years, who was a favourite of my grandfather and my mother, and hence whose work I'd been given, yet whose sufferings I'd somehow not grasped. Emily Dickinson wrote, in a famous letter to her Uncle Sweetser: 'Much has occurred, dear Uncle, since my writing you – so much – that I stagger as I write, in its sharp remembrance. Summers of bloom – and months of frost, and days of jingling bells, yet all the while this hand upon our fireside. Today has been so glad without, and yet so grieved within . . . I cannot always see the light – please tell me if it shines.'[22]

The hand upon the fireside has been taken by biographers as a delicate allusion to a depressive state, and there are further clues in the poet's reclusive life. When Dickinson was an adolescent, she wrote to a friend that she was being kept out of school on health grounds, because of her 'low spirits'. In 1846, when she was fifteen, she wrote to another friend, Abiah Root, and mentioned an attack of severe melancholy that had occurred after the death two years previously of her friend Sophia Holland: 'I shed no tear, for my heart was too full to weep, but after she was laid in her coffin & I felt I could not call her back again I gave way to a fixed melancholy. I told no one the cause of my grief, though it was gnawing at my very heart strings.'[23]

A year on from that letter, aged sixteen, Dickinson moved to the Mount Holyoke Female Seminary at South Hadley, about ten miles from her home in Amherst, Massachusetts. Dickinson's year there was unhappy and damaging; it coincided with a determined effort at religious conversion. As the novelist Anna Mary Wells describes it:

> This [year of study] was not a success either therapeutically or pedagogically. Mount Holyoke had an excellent academic reputation, but during the year Emily Dickinson was there its director and faculty were primarily interested in a heavily charged emotional effort to persuade all those students who were not yet professing Christians to 'give themselves to Christ'. Emily was one of very few who were able to hold out for the year against the overwhelmingly dominant personality of Mary Lyon, but she did it at the price of great personal unhappiness and anxiety.[24]

Again, the circumstantial evidence is strong that Dickinson was not just unhappy but depressed. Her condition was diagnosed as homesickness and treated by an enforced return home and a six-week recuperation there. In her work thereafter there are continual hints, expressed with characteristic cryptic brevity, of the nature of her mental torment. Her poetry is famously allusive and it's important not to read into her words a direct expression of her state of mind. She explores, with often startling imagery and imperfect semi-rhymes, ways of seeing the world – and they may not be hers. We are no more able to infer the poet's intellectual autobiography in Emily Dickinson's case than we are the life story of Shakespeare from his sonnets (though many have fruitlessly tried). We may not do it. Yet the compressed fragments of her poetry express what I felt, in words that I'd read but whose unerring accuracy I'd not previously perceived.[25]

There is a pain – so utter –
It swallows substance up –
Then covers the Abyss with Trance –
So Memory can step –
Around – across – upon it –
As One within a Swoon –
Goes safely – where an open eye –
Would drop Him – Bone by Bone. [599]

The pain of depression is the most intense I've felt. But it's more than that. It's 'utter'. It takes over everything, to the extent that 'it swallows substance up'. There is nothing left of life. The bluntness of 'utter' captures it. The world is a desolate space with no joy and no leaven. Here it is, in the sparsest yet completest of descriptions. I was knowing this in the way, exactly the way, in which she encapsulated it. Nor in depression is there any prospect of alleviating the pain. There is no recovery or means of breaking down what the poet imagines as an impregnable fortress.

Me from Myself – to banish –
Had I Art –
Invincible my Fortress
Unto All Heart –

But since Myself – assault Me –
How have I peace
Except by subjugating
Consciousness?

And since We're mutual Monarch
How this be
Except by Abdication –
Me – of Me? [642]

If she didn't, after all, suffer severe depression herself, she understood and expressed with pellucidity its nature. Very possibly, she knew it intimately and in an unbroken stretch of years, even an entire adult lifetime. For when you're in a depressive state, it's your own thoughts that are the source of anguish. They assault you; you assault you. And because this embattlement is the unyielding product of your consciousness, it will cease only when that conscious state is ended by death – by me giving up the struggle and leaving 'me'. There's an implied dualism here, of 'mutual' sovereignty over the self, yet how percipiently does the poet imagine, in accord with all we now know of the physiological origins of consciousness, the identity of the mind with the person. She can't separate these entities; she hence can never escape the weight of her despair.

The same thought occurs about a year later in Emily Dickinson's output (written in 1863 or thereabouts). It's conveyed simultaneously delphically, yet with the brute force of explicit reference to the end of life.

No drug for Consciousness – can be –
Alternative to die
Is Nature's only Pharmacy
For Being's Malady – [from 786]

There is no medicine, no drug, that can heal the poet's torment. Only death can bring deliverance. And what's most striking about the stanza (which, with the unconventional final dash, is the conclusion of the poem) is the final noun phrase, 'Being's Malady'. Throughout this book I've referred to *mental* illness and *mental* disorder, while stressing that the distinction commonly made – in medical practice, and not only in normal speech – between diseases of the mind and those of the body is artificial. The model of Cartesian dualism does not describe how humans really are, but we're stuck with its terminology. Emily Dickinson gets beyond it, not only as a powerful turn of speech but as a clinical description. The tortured mind is *being's* malady. For her, depression isn't a state of downcast spirits, a sickness of the soul, but an agony that is literally existential; its end comes only with our own.

For the poet there was no pharmacy other than the inevitability of mortality. The sole assurance of release was the peace of the grave. For us, counterposed against that private world of mental disorder, working to understand the condition and dispel it, there is the public practice of modern science. It has made immense progress in understanding the disorders of the mind and devising effective drugs and treatments. Let me summarise, finally, what we know and also what we have yet to discover.

Science and the depressive state

Though depression, this distinctive malady of being, has been part of the human condition for at least as long as civilisation and probably for much longer, it is comprehensible to modern science. There is every possibility already not only of effective intervention to hasten and secure remission from depression, but to cure it – not for everyone, but for many. Depression will prove increasingly soluble, not because modern science is anywhere close to understanding the nature of consciousness and how its workings can break down, but because we know in principle the methods by which these questions can be investigated and their answers discovered. Just as the science of the brain has made great advances through various interconnected disciplines, so the treatment of depression and other disorders has become grounded in evidence rather than superstition. The evidence we can draw on is randomised clinical trials of psychological and pharmacological treatments. The methods that yield these conclusions are derived from the practice of inductive reasoning. We owe its methods in the first place to the Hellenistic civilisation of Hippocrates and Aristotle. Science isn't inevitable in the human condition; it's what we are obliged to defend.

There are other treatments for depression and mental disorder that I've discussed, where the evidence is slighter or the dangers greater. Some of these are widespread but questionable (such as psychodynamic therapy) and some are reserved for extreme cases because there are potentially severe side effects (such as ECT, though it is a safer and more humane procedure than is depicted in popular culture). For the patient, severe depression is devastating but treatable. For the scientist, doctor or psychologist, it's in many ways mysterious but in principle comprehensible. Depression is a tractable area of intellectual inquiry and discovery.

The great mystery of the workings of consciousness, and how those workings can break down without warning or apparent cause, remains to be explained. It's unlike the human ailments that we (somewhat arbitrarily) classify as physical, because we don't have direct access to the conditions that create such suffering. To gain insight into the nature of depression and psychiatric disorders, all we have available for direct observation are the sufferer and what they recount. Brain-scanning techniques have advanced rapidly over a generation and they potentially can tell us something about how the mental processes of a sick patient differ from those of a healthy person. But such work is only suggestive at this point. In the main, the material available that enables us to partially understand the mental universe of severe depression is how sufferers describe it.

Routes to knowledge

Hence there is a valuable – an essential – source of knowledge about mental disorder drawn from beyond the scientific study of the brain. It's this body of reflection in fields including poetry, art, history and anthropology. It comprises the testimony of those who've experienced depression and, whether for personal release or to enhance others' understanding, been determined to describe the experience. We don't yet know much about the neural correlates of consciousness; the technology is not advanced enough, and perhaps this will forever be a problem without a solution. So the description of mental disorder, along with its depiction in works of the imagination, is the best evidence we yet have of what it *is*, and not only what it *feels like*. Because the affective disorders of humanity are known to us principally by their symptoms and only speculatively (if at all) by their causes, it's to the sufferers that we can only look to make them plain. The cry across ages from those who know depression is Macbeth's entreaty:

> Canst thou not minister to a mind diseased,
> Pluck from the memory a rooted sorrow,
> Raze out the written troubles of the brain,
> And with some sweet oblivious antidote
> Cleanse the fraught bosom of that perilous stuff
> Which weighs upon the heart?[26]

This is the type of knowledge that art and literature provide. It concerns the world not of measurable fact but of feeling and meaning. It serves as an intermediary between how we experience the world privately and how we understand it publicly. It's a concept I take from the literary critic F. R. Leavis, who referred to 'the type of all judgments and valuations belonging to what in my unphilosophical way I've formed the habit of calling the "third realm" – the collaboratively created human world, the realm of what is *neither* public in the sense belonging to science (it can't be weighed or tripped over or brought into the laboratory or pointed to) nor merely private and personal (consider the nature of a language, of the language we can't do without – and literature is a manifestation of language)'.[27]

Art and literature are located in this third realm. The way they are understood is through a collaboration between creator and audience, and the experience of them is not the same for everyone. In this they are unlike scientific discovery, which is a body of evidence that, though always open to being disproved, holds consistently across cultures, epochs and audiences. Great writers cause us to reflect on the puzzles and dilemmas of human existence, and to do so without any natural limit. They don't instruct us in the 'correct' answers, as lesser and didactic writers might.

Take Robert Browning, in a poem that spoke to me through many years of my perplexity at failing in the practice of life, and does so now in a slightly different way. In 'Andrea del Sarto', a painter tells of how his art is on the face of it perfectly executed and how effortless he finds the task ('At any rate 'tis easy, all of it!'), yet he cannot ultimately succeed for his work is bloodless. He lacks any vision of his own, and can only deploy 'this low-pulsed forthright craftsman's hand of mine'. It's others, less ostensibly accomplished, who 'reach many a time a heaven that's shut to me'. And he observes, in lines that are much quoted: 'Ah, but a man's reach should exceed his grasp, / Or what's a heaven for?'[28]

In life, we require a challenge for us to be fulfilled; our reach, our ambitions should outstrip what we are immediately capable of. And we strive. Many people will recognise Browning's lines while not being aware of the source, which is indeed my point about their being part of the intermediate world, the third realm. They express an enduring truth about the human condition in a way that we constantly interpret and reinterpret in our own lives, and they do so in language that is immediate, succinct and memorable. It's what great literature does.

I carried these lines, and this conundrum, with me for years and decades. I strove and failed, and the failure weighed on me ever more heavily. I had the superficial accomplishments of professional recognition. Sustaining this was technical proficiency, like the painter's, but the outcome was hollow. The true extent of my grasp, in personal and public life, seemed ever more constrained. The only way to narrow that gap was to accept that I was incapable of greater things, and to live with the knowledge of my failings. And as years passed, the failings became ever more salient in my imagination. For me now, after the depression, I have learnt to see the details slightly differently, but still with the knowledge that fulfilment requires achieving something beyond me. Psychologically, I'm less prone to the catastrophising beliefs about what I had achieved relative to what I sought, and in particular what others expected of me. The demoralised state where I surveyed what I had achieved and saw only wreckage has been eased. I know that the lives even of those I admire and love the most are chequered too, if doubtless less so than mine, and I'm not fated to forever disappoint those who look to me. To fall short is no dishonour. It still profits a person to strive, to seek and ultimately not to find.

Depression and the intermediate world

The experience of mental disorder is part of that intermediate world too. It's in the mind of the sufferer, and it takes over the way you perceive life. But because of human ingenuity and the capacity for sympathy, the confinement need not be suffered alone. The burden can be spread, mitigated and lifted. Scientific research has yielded remedies and is striving for knowledge of why depressive disorders occur and how they can be more effectively treated. And between the private experience of the sufferer and the public practice of science is the province of shared feeling, intuition and affinity – things that we create collectively and collaboratively, and that thereby gain their meaning.

It brings us back to language, which Leavis rightly cites as an example of the intermediate world. He might have put it more strongly: language is the *quintessential* part of this realm. It is the preeminent and most distinctive of human faculties, for it's shared by all of our species and by no other. We are thinking beings, but we don't think in the words of our language. Rather, in the view of modern cognitive science, we think in a language of thought – a representation of concepts, entities, events and properties in the brain. Pinker refers to these symbols as a universal mentalese, and says: 'Knowing a language . . . is knowing how to translate mentalese into strings of words and vice-versa. People without a language would still have mentalese, and babies and many nonhuman animals presumably have simpler dialects. Indeed, if babies did not have a mentalese to translate to and from English, it is not clear how learning English could take place, or even what learning English could mean.'[29]

The activities of the mind and of perception are unique to us individually, but the way we interact with the world is not arbitrary. Pinker suggests that the languages of thought, the mentalese, are the same. The ability to translate mentalese into English means knowing the collaboratively constructed conventions under which some strings of English words are grammatical and others not, and that words have certain tacitly understood meanings that are shared by other English speakers. Recall the way Dr Johnson compiled his dictionary, as all lexicographers since have done. His method was not to issue arbitrary edicts about definitions but to investigate how words were used in the world.

When the mind breaks down, the habits of this shared realm of socially constructed meanings help call us back. When I lost hold of my senses, medical science and clinical psychology stopped my fall and explained to me what was happening. The ascent to sanity was not smooth, but every time I fell it was to a slightly higher plateau. I began once more to look in the world of letters and art for fixed points of human experience that I could recognise and hold to, so that I could take my place amid the familiar rather than stumble in a wilderness. And eventually I did more than resume normality: I attained that desired state of a 'new normal', where emotions are sharper and understanding of those around me greater. The catastrophising instinct, where setbacks become disasters and doubt is turned to self-hatred, was tamed.

Colour in shadow

I could not properly read again for a long time, in the sense of immersing myself in a book for hours at a stretch. But I would visit galleries: the principal ones in central London, the National Gallery, the Tate and the Courtauld, so that I could lose myself in just a single room or corridor at a time, without feeling conspicuous if I either sat and looked vacant for an hour or if I left again in anxiety within a few minutes.

It will sound trite if I claim that paintings of bucolic idylls alleviate depressed moods and those of happy subjects dispel the darkness, and unfortunately it isn't true (or at least it wasn't for me). But I noticed something I'd previously known intellectually but not truly felt. I went to the paintings, in oil and watercolour, of J. M. W. Turner, owing to his reputation as 'the painter of light', hoping they would – by osmosis, perhaps – bring me out of the darkness of depression. It was naive, but not completely fanciful. John Ruskin, of all critics, noted that Turner after 1820 became 'without rival, the painter of the loveliness and light of creation'. Ruskin observed that even in so early a work as *Apollo and Python*, painted in 1811 and depicting (after the ancient Greek poet Callimachus) the sun god Apollo killing a giant dragon, Turner depicted 'rose colour and blue on the clouds, as well as gold'. It's in Tate Britain. As he progressed in his mastery of colour, Turner arrived at the 'colour chord' using scarlet. No one before had thought to depict shadow in scarlet tones, rather than as black, in counterpoint to the light of the sun.[30]

It will seem a bathetic observation by comparison, but I looked at the shadow in these works (a much later one, held by the Courtauld, is *Dawn After the Wreck*, painted around 1841) and found that it was not black as I'd supposed. It was, as Ruskin said of this discovery of scarlet shadow, 'the glory of sunshine'. I took it as a metaphor for the blackness of depression. It is actually, even in its darkest times, not black but part of the range of colour. An artist and a critic of unrivalled perception had yielded to me that discovery of the natural world, their perception of colour and light, and I assimilated it into the world of my imagination.

Not black. No longer black. Even when my depressive state reasserted itself and took me back to a deep trough, I could gauge it was not my lowest point. It was not black but merely a shade of grey. A lesser cultural observer than Ruskin described this too in a way that I carried with me. G. K. Chesterton is a minor literary figure of the early twentieth century whose memory is kept alive principally owing to his detective stories about the cleric Father Brown and through the efforts of devotees of his own Catholic orthodoxy. His literary work is marred by an ostentatious commitment to extravagant paradox, and his social criticism is debased by anti-Semitism.[31] Yet there are moments of acute insight. One such is in an essay about greyness, rooted in stereotypical observations about the inclemency of the English climate but concluding with this flourish: 'Grey is a colour that always seems on the eve of changing to some other colour; of brightening into blue or blanching into white or bursting into green and gold. So we may be perpetually reminded of the indefinite hope that is in doubt itself; and when there is grey weather in our hills or grey hairs in our heads, perhaps they may still remind us of the morning.'[32]

EPILOGUE

THE END OF DEPRESSION

The despair of depression incrementally gave way to mere drabness. With these varying shades of grey came the dissolution of mental disorder and its pacification by more benign and creative ways of thought. And thereafter I encountered the possibility of change and even of rejuvenation: not a return to the normal condition of life, which I had navigated while never mastering, but the attainment of a 'new normal' state of contentment without complacency.

Entering once more the realm of works of the imagination, of literature, art and music, is (it was for me) a means of taming the terrors of loneliness and acclimatising the mind to a better way of thinking. There was no defining moment when I was cured, but the difference was not one merely of degree, nor was I exactly the same person as before. It took me two attempts to come off medication before I felt safe to permanently discontinue it. Though the principles of psychological therapy were carefully and gently impressed upon me so that I could practise them wherever I was and at any time, I found that I needed the assurance and expertise of a clinical psychologist to rehearse them again, in a second series of sessions. And then I let go. I measured progress by achieving things that previously had been beyond me. I could open the front door going in either direction without having to mentally build up to it. I could read for hours rather than minutes at a stretch. The doubts remained but the turmoil was stilled. Some things never returned; the edge seems to have gone from my memory, but it's a minor inconvenience.

On our first meeting, when I could barely stand and scarcely speak, the mental-health campaigner Lord Stevenson recounted a comment he'd heard from Alastair Campbell, a sufferer like me though with a more public role: depression is akin to being in a trench and you have no idea how to get out; and when you're out, you have no idea how you ever fell in. Depression remains a harrowing affliction with immense and usually unexpressed costs in the quality, even the continuation, of people's lives. It has been this way for millennia. It was known by our forebears in identical form though with different terminology. Burton wrote in his *Anatomy* that 'fear and sorrow are the true characters and inseparable companions of melancholy', which strikes down the sufferer 'without any apparent occasion'.[1]

That's what it's like. Depression is a sudden transport to a landscape of incapacitating strangeness, where rescuers blessedly came to find me when I cried out. One day, probably not in our lifetimes but perhaps sooner than we think, medical science will be able to map the contours of that barren terrain and give hope to millions of others, including those who have languished there for a lifetime without believing there is a possibility of escape. The surest, swiftest path to achieving that goal is to impress upon sufferers and policymakers that severe clinical depression yields to investigation, that it is in principle curable, and that those whose lives are devastated by this disorder are not responsible for their plight.

Acknowledgements

I'm indebted to Dr Jeremy Nathan, Dr Annemarie O'Connor and Lord (Dennis) Stevenson.

Professor Matthew Broome and Dr John Williams lent their dual expertise on mental health, neuroscience, neuropharmacology and psychiatry. They read the manuscript in draft, corrected errors of fact and interpretation, and much expanded my knowledge of their specialist fields. Professor Bob Borsley, who has long been my patient guide to debates in theoretical linguistics, read the sections dealing with language. Whatever mistakes remain in the book are my responsibility alone.

Jenny Lord of Weidenfeld & Nicolson made extensive and invaluable suggestions on how to improve the book and render its argument more transparent. I'm grateful to her for publishing it and to Will Francis of Janklow & Nesbit, my agent, for proposing it.

Nicola Jeal, Editor of *The Times* magazine, urged me to write an essay on my experience of depression, which garnered a huge response from readers and was the initial impetus for this book.

James Harding, John Witherow, Keith Blackmore, Emma Tucker and Tony Gallagher, successive Editors and

Deputy Editors of *The Times*, have long given me encouragement and support, not least during the period described in this book. I'm also thankful for many other valued colleagues, past and present, including David Aaronovitch, Ian Brunskill, Jessica Carsen, Philip Collins, Daniel Finkelstein, Robbie Millen, Simon Nixon, Matthew Parris, Hugo Rifkind, Beth Rigby and Giles Whittell.

Gratitude finally goes to Aisha Ali-Khan, Katherine Arora, Andrew Bailey, Catherine Bell, Martin Bell, Melissa Bell, Sharon Bell, Professor Patricia Clavin, Nick Cohen, Myrto Gelati, Francesca Gonshaw, Nigel Hankin, Gina Higgins, Edina Hodzic, Alexandra Kamm, Eileen Kamm, Eleanor Kamm, Richard Kamm, Anna Mandoki, Beth McCann, Peter Oppenheimer, Agnès Poirier, Professor Geoffrey Pullum, John Rentoul, Professor Sarah Savant, Elise Valmorbida, Francis Wheen and Arani Yogadeva.

Notes

Preface

1 Aaron Beck, *Cognitive Therapy and the Emotional Disorders*, London: Penguin, 1979, p. 26.
2 Elke Van Hoof, 'Lockdown is the world's biggest psychological experiment – and we will pay the price', World Economic Forum, 9 April 2020, available at: https://www.weforum.org/agenda/2020/04/this-is-the-psychological-side-of-the-covid-19-pandemic-that-were-ignoring/
3 Public Health England, Guidance for the Public on the Mental Health and Wellbeing Aspects of Coronavirus (COVID-19), updated 31 March 2020, available at: https://www.gov.uk/government/publications/covid-19-guidance-for-the-public-on-mental-health-and-wellbeing/guidance-for-the-public-on-the-mental-health-and-wellbeing-aspects-of-coronavirus-covid-19

Chapter 1

1 Letter to John Stuart, 23 January 1841, in *The Collected Works of Abraham Lincoln*, edited by Roy P. Basler, 8 vols, vol. 1, New Brunswick, NJ: Rutgers University Press, 1953, pp. 229–30.
2 Biographers have traditionally interpreted this episode in Lincoln's life in the context of the end of his engagement to Mary Todd in 1841, and his estrangement from her till the autumn of the following year. More recent writers have seen

it as evidence of a long struggle with clinical depression. See Joshua Wolf Shenk, *Lincoln's Melancholy: How Depression Challenged a President and Fueled his Greatness*, Boston, MA: Houghton Mifflin, 2005. Judged by the character of her grieving, Mary Todd Lincoln (as she became) almost certainly suffered mental illness after her husband's assassination, compounded by debts. Her son Robert had her committed to a mental institution in 1875 and she struggled to clear her name of the stigma of insanity.

3 World Health Organization, *Factsheet on Depression*, 30 January 2020, available at: https://www.who.int/mental_health/management/depression/en/

4 World Health Organization, *Depression and Other Common Mental Health Disorders: Global Health Estimates*, 2017, p. 5, available at: https://apps.who.int/iris/bitstream/handle/10665/254610/WHOMSD?sequence=1

5 T. Vos, R. M. Barber, B. Bell et al., 'Global, regional, and national incidence, prevalence, and years lived with disability for 301 acute and chronic diseases and injuries in 188 countries, 1990–2013: A systematic analysis for the Global Burden of Disease study', *The Lancet*, 386 (9995), 2013, pp. 743–800.

6 A. J. Ferrari, F. J. Charlson, R. E. Norman et al., 'Burden of depressive disorders by country, sex, age, and year: Findings from the Global Burden of Disease study 2010', *PLOS Medicine*, 10 (11), 2013.

7 J. Evans, I. Macrory and C. Randall, 'Measuring national wellbeing: Life in the UK, 2016', Office for National Statistics, 2016, available at: https://www.ons.gov.uk/peoplepopulationandcommunity/wellbeing/articles/measuringnationalwellbeing/2016#how-good-is-our-health

8 The NICE definitions of these symptoms are set out at: https://www.nice.org.uk/guidance/cg90/chapter/Appendix-Assessing-depression-and-its-severity

9 World Health Organization, op cit., 2017, p. 5.

Chapter 2: What Causes Depression?

1 Hippocrates, 'The Sacred Disease', from *The Medical Works of Hippocrates*, translated by John Chadwick and W. N. Mann, Springfield, IL: Charles C. Thomas, 1950, included in the anthology *Medicine and Western Civilization*, edited by David A. Rothman, Steven Marcus and Stephanie A. Kiceluk, New Brunswick, NJ: Rutgers University Press, 1995, p. 143.

2 Michael Oakeshott, 'On being conservative', *Rationalism in Politics and Other Essays*, London: Methuen & Co. Ltd, 1962, pp. 436–7.

3 Christopher Hitchens, 'Topic of Cancer', *Vanity Fair*, September 2010. It's reproduced in his posthumous book *Mortality*, London: Atlantic Books, 2012, p. 6.

4 Cited in Geddes MacGregor, *Images of Afterlife: Beliefs from Antiquity to Modern Times*, New York: Paragon House, 1992, p. 177.

5 David G. Blanchflower and Andrew J. Oswald, 'Is well-being U-shaped over the life cycle?', *Social Science & Medicine* 66, 2008, pp. 1733-1749. The paper, whose findings Blanchflower has since replicated for 145 advanced and developing countries, is available at: https://www.dartmouth.edu/~blnchflr/papers/welbbeingssm.pdf

6 Ibid., p. 1735.

7 A full and illuminating study of this entire question is Jonathan Rauch, *The Happiness Curve: Why Life Gets Better After Midlife*, London: Bloomsbury, 2018. Rauch, whose writings on social issues I've long admired, had the exact experience I did of encountering Cole's paintings at a perplexing time in life. I was gratified that it wasn't just me who'd looked at them and found them apposite, and I've drawn on his excellent programmatic description of them.

8 W. B. Yeats, *The Poems*, London: J. M. Dent & Sons, 1990, p. 243.

9 There again, Yeats got other compensations. As Maud Gonne, the object of his ardour, famously replied to his proposal of marriage in 1902: 'You make beautiful poetry out of what you call your unhappiness and you are happy in that. Marriage would be such a dull affair. Poets should never marry.' Quoted in Bernard O'Donoghue, 'Yeats the love poet', *Yeats Annual No. 20*, 2016, p. 101.

10 David Clark and Richard Layard, *Thrive: The Revolutionary Potential of Evidence-Based Psychological Therapies*, London: Penguin, 2014, p. 122.

11 Hippocrates, 'The Sacred Disease', op cit., p. 143.

12 This account draws on Stanley Finger and Hiran R. Fernando, 'E. George Squier and the discovery of cranial trepanation: A landmark in the history of surgery and ancient medicine', *Journal of the History of Medicine and Allied Sciences*, 56 (4), October 2001, pp. 353–81. A full treatment of the subject is John W. Verano, *Holes in the Head: The Art and Archaeology of Trepanation in Ancient Peru*, Washington DC: Dumbarton Oaks Research Library for Harvard University, 2016.

13 This is a fault of fundamentalists quite as much as liberal theologians, by seeking to reduce accounts of the miraculous to 'non-miraculous but scientifically plausible connections, all of them quite absent from the biblical text'. I owe this point to the Hebrew scholar James Barr. See his *Fundamentalism*, London: SCM Press, 1977, especially pp. 234–59 for an exposition of this oddity. The quoted words are on p. 242.

14 Mark 5: 2–5. The quotation and the one following are from the Authorised Version of the Bible of 1611.

15 Carlos Espí Forcén and Fernando Espí Forcén, 'Demonic possessions and mental illness: Discussion of selected cases in late medieval hagiographical literature', *Early Science and Medicine*, 19 (3), 2014, p. 262.

16 H. R. Trevor-Roper, *The European Witch-Craze of the 16th and 17th Centuries*, London: Penguin, 1969, p. 116.

17 Keith Thomas, *Religion and the Decline of Magic: Studies in Popular Beliefs in Sixteenth- and Seventeenth-Century England*, London: Weidenfeld & Nicolson, 1971, p. 16.
18 Michael MacDonald, *Mystical Bedlam: Madness, Anxiety and Healing in Seventeenth-Century England*, Cambridge: Cambridge University Press, 2009.
19 Ibid., p. 213.
20 Andrew Scull, *Madness in Civilization: A Cultural History of Insanity from the Bible to Freud, from the Madhouse to Modern Medicine*, London: Thames and Hudson, 2016, p. 96.
21 'I do set my bow in the cloud, and it shall be for a token of a covenant between me and the earth.' Genesis 9:13.
22 Isaac Newton, *Opticks*, Book 3, Part 1, question 31, cited in Paul D. Schweizer, 'John Constable, rainbow science, and English color theory', *The Art Bulletin*, 64 (3), September 1982, p. 444.
23 'But were this world ever so perfect a production, it must still remain uncertain whether all the excellences of the work can justly be ascribed to the workman.' David Hume, *Dialogues Concerning Natural Religion*, London: Penguin Classics, 1990 (first published 1779), p. 77.
24 Martin Gardner, *When You Were a Tadpole and I Was a Fish: And Other Speculations About This and That*, New York: Hill and Wang, 2009, p. 9. I've taken the information about Newton's fascination with numerology and alchemy from Gardner's essay in this volume 'Isaac Newton's Vast Ocean of Truth', pp. 9–13.
25 Graham C. L. Davey, '"Spirit Possession" and Mental Health', *Psychology Today*, 31 December 2014, available at: https://www.psychologytoday.com/gb/blog/why-we-worry/201412/spirit-possession-and-mental-health
26 'Christian GP who performed exorcism on patient is struck off', National Secular Society, 29 January 2015, available at: https://www.secularism.org.uk/news/2015/01/

christian-gp-who-performed-exorcism-on-patient-is-struck-off

27 Richard Gallagher, 'As a psychiatrist, I diagnose mental illness. Also, I help spot demonic possession', *Washington Post*, 1 July 2016.

28 This case is set out at length by Homayun Sidky, *Witchcraft, Lycanthropy, Drugs and Disease: An Anthropological Study of the European Witch-Hunts*, New York: Peter Lang, 1997.

29 Francis Crick, *The Astonishing Hypothesis: The Scientific Search for the Soul*, London: Simon & Schuster, 1994, p. 3.

30 The philosophical premises in Crick's book have been criticised by John Searle, a leading philosopher of mind, but Searle urges even so that 'when you read the book you can ignore the philosophical parts and just learn about the psychology of vision and brain science', John Searle, *The Mystery of Consciousness*, London: Granta Publications, 1997, p. 35. So that's what I've done.

31 James Boswell, *The Life of Samuel Johnson*, London: Penguin Classics, 2008 (first published 1791), p. 365.

32 The polling evidence is set out in a factsheet by the independent Religion Media Centre, available at: https://religionmediacentre.org.uk/factsheets/secularisation-in-britain/

33 Steven Pinker, *How the Mind Works*, London: Penguin, 1998, p. 77.

34 I owe this observation to Lilli Alanen, 'Reconsidering Descartes's notion of the mind-body union', *Synthese*, 106 (1), January 1996, p. 3.

35 The critique is set out in Gilbert Ryle, *The Concept of Mind*, London: Hutchinson, 1949, especially pp. 14–15 and 154–5.

36 Thomas De Quincey, *Confessions of an English Opium-Eater and Other Writings*, Oxford: Oxford World's Classics, 2013, p. 72 (first published 1821).

37 The experiment is recounted in O. Blanke, S. Ortigue, T. Landis et al., 'Stimulating illusory own-body perceptions', *Nature*, 419, 1992, pp. 269–70. I've taken the reference,

and the information on the effects of cerebral anoxia, from Terence Hines, *Pseudoscience and the Paranormal*, New York: Prometheus Books, 2003, pp. 103-05.

38 Kendler sets out a philosophical framework for psychiatry in 'Toward a Philosophical Structure for Psychiatry', *American Journal of Psychiatry*; 162 (3), March 2005, pp. 433-440. The case for 'metaphysical ecumenism' is advanced by George Graham, *The Disordered Mind: An Introduction to Philosophy of Mind and Mental Illness*, Abingdon: Routledge, 2013 (2nd edn), p. 92.

39 A fascinating study of the history of phrenology in Britain, from which I've taken the Whately reference, is T. M. Parssinen, 'Popular science and society: The phrenology movement in early Victorian Britain', *Journal of Social History*, 8 (1), Autumn 1974, pp. 1–20.

40 The cerebral cortex, which is the outermost portion of the brain, is divided into four lobes: the frontal lobe, at the front of the brain; the parietal lobe, in the middle; the temporal lobe, at the bottom; and the occipital lobe, at the back.

41 Daniel Mirman, Qi Chen, Yongsheng Zhang, Ze Wang, Olufunsho K. Faseyitan, H. Branch Coslett and Myrna F. Schwartz, 'Neural organization of spoken language revealed by lesion-symptom mapping', *Nature Communications*, 6, article number 6762, 2015, available at: https://www.ncbi.nlm.nih.gov/pmc/articles/PMC4400840/

42 Wallace defended an American medium called Henry Slade even after Slade had been caught in subterfuge at a seance and was being prosecuted for fraud. A letter from Wallace on this case from 1876 is reproduced by the Alfred Russel Wallace Correspondence Project at http://wallaceletters.info/content/spiritualism. For an excellent account of Wallace's thinking on science, politics (he became a socialist) and spiritualism, and his belief in the consilience of all these, see D. A. Stack, 'The first Darwinian Left: Radical and Socialist responses to Darwin,

1859–1914', *History of Political Thought*, 21 (4), Winter 2000, pp. 682–710.

43 This was for a booklet accompanying a special issue of Royal Mail stamps to mark the centenary of the Nobel Prizes in 2001. The full article is available at Josephson's home page at the Physics Department of Cambridge University: http://www.tcm.phy.cam.ac.uk/~bdj10/stamps/text.html

44 This sort of special pleading was satirised by Robert Browning in his poem 'Mr Sludge – "The Medium"', available in many editions but mine is Robert Browning, *The Poems: Volume 1*, London: Penguin Classics, 1993, pp. 821–60. The model for the fraudulent Mr Sludge was the celebrated spiritualist medium Daniel Dunglas Home (1833–86), who was credited among other gifts with the power of levitation: he was reported at one seance to have floated out of a third-floor window and returned by the same route. Because he was never actually detected in cheating, he is still treated by some writers as if his feats may have been genuine. A book-length example of this curious genre is Peter Lamont, *The First Psychic: The Peculiar Mystery of a Victorian Wizard*, London: Little, Brown, 2006. There is in fact no 'mystery' that Home used conjuring tricks for his effects.

45 Daniel Dennett, *Darwin's Dangerous Idea: Evolution and the Meanings of Life*, London: Penguin, 1995, pp. 511–12.

46 Though we don't know precisely how antidepressants work, this is not to devalue them or dispute their effectiveness. We don't know, either, exactly how paracetamol works to relieve headaches and other pains. A popular but not definitive explanation is that it does so by reducing the production of prostaglandins in the brain and spinal cord. Perhaps because sceptics have less at stake in the matter, there is no public outcry at the widespread use of these non-prescription pills.

47 Stephen Leacock, Preface to *Sunshine Sketches of a Little Town*,

London: Prion Books, 2000 (original publication 1912), p. xvii.

48 The best explanation I know of the financial crisis, its historical antecedents and its economic consequences is Barry Eichengreen, *Hall of Mirrors: The Great Depression, the Great Recession and the Uses – and Misuses – of History*, Oxford: Oxford University Press, 2015. Among the factors causing the collapse was, says Eichengreen, 'the naïve belief that policy had tamed the [business] cycle . . . Encouraged by the belief that sharp swings in economic activity were no more, commercial banks used more leverage. Investors took more risk' (p. 3).

49 Edward Bullmore, *The Inflamed Mind: A Radical New Approach to Depression*, London: Short Books, 2018, p. 179.

50 Ibid., p. 24.

Chapter 3: How We Understand Depression

1 William Styron, 'Why Primo Levi need not have died', *New York Times*, 19 December 1988.

2 *King Lear*, Act 4, Scene 5. The edition I've used for all Shakespeare quotations is *The Oxford Shakespeare: The Complete Works*, edited by John Jowett, William Montgomery, Gary Taylor and Stanley Wells, Oxford: Oxford University Press, 2005 (2nd edn).

3 The dangers of inferring Shakespearian biography from works of the imagination are exemplified in the myth, which grew up in the mid nineteenth century, that Shakespeare was merely the pseudonym or frontman of a writer of noble birth. James Shapiro, in *Contested Will: Who Wrote Shakespeare?*, London: Faber & Faber, 2010, has brilliantly traced the entirely fallacious belief of alternative authorship back to this conceptual error that the works must be autobiographical.

4 A. C. Bradley, *Shakespearean Tragedy: Lectures on Hamlet, Othello, King Lear, and Macbeth*, London: Penguin, 1991, p. 121 (first published 1904).

5 The mental disorder of Hamlet is implicit in T. S. Eliot's influential critique that the play is an artistic failure because Hamlet's hostility to his mother exceeds any cause in the known facts. That's what a depressive disorder is like, though: it's an abnormal sadness. And even then, the facts external to the play are ample to explain (not excuse) his behaviour to his mother and Ophelia. This last point is the brisk response of John Dover Wilson in 'Mr T. S. Eliot's Theory of Hamlet', in *What Happens in Hamlet*, Cambridge: Cambridge University Press, 1951 (3rd edn), pp. 305–8. I've also gained a lot in understanding the psychoanalytic interpretation of the play from William Kerrigan's short book *Hamlet's Perfection*, Baltimore, MD: Johns Hopkins University Press, 1994. Among other things, the book told me that it was to resolve the puzzle of Hamlet's delay that Freud first invoked the concept of the Oedipus complex (in a letter to Wilhelm Fliess in 1897).

6 William Styron, *Darkness Visible: A Memoir of Madness*, London: Jonathan Cape, 1991, p.5.

7 Francis Bacon, 'Of Death' (1612), in *The Essays*, London: Penguin Classics, 1985, p. 64. He's alluding to Lucretius, *De Rerum Natura*, who in Book III discusses the mortality of the soul.

8 Arthur Schopenhauer, translated by E. F. J. Payne, *The World as Will and Representation*, 2 vols, vol. 2, Indian Hills, CO: Falcon's Wing Press, 1958, p. 628.

9 Letter in *The American Scholar*, 59 (2), Spring 1990, pp. 319–20.

10 Primo Levi, translated by Raymond Rosenthal, *The Drowned and the Saved*, London: Abacus, 1989, p. 52.

11 The suicide note is held by the National Library of Israel and can be read in facsimile (in German) and in translation at: http://web.nli.org.il/sites/NLI/English/collections/personalsites/archive_treasures/Pages/stefan-zweig.aspx

12 Tom Burns, *Our Necessary Shadow: The Nature and Meaning of Psychiatry*, London: Penguin, 2013, p. xxxii.
13 Edward Shorter in Ingrid Wickelgren, 'Trouble at the heart of psychiatry's revised rule book', *Scientific American* blog, 9 May 2012. Shorter has set out his critique at length in *How Everyone Became Depressed: The Rise and Fall of the Nervous Breakdown*, New York: Oxford University Press, 2013.
14 Burns, op. cit., p. xxxii.
15 Oliver Sacks, *The Man Who Mistook His Wife for a Hat*, London: Gerald Duckworth & Co. Ltd, 1985, p. 105.
16 My account of the characteristics of depression as an affective disorder draws on Constance Hammen and Edward Watkins, *Depression*, Abingdon: Routledge, 2018 (3rd edn), Chapter 1.
17 Samuel Johnson, *The History of Rasselas, Prince of Abissinia*, London: Penguin Classics, 1985 (first published 1759), p. 80.
18 Ibid., p. 81.
19 Cited in Carey McIntosh, 'Johnson's debate with Stoicism', *ELH*, 33 (3), September 1966, p. 330.
20 Evelyn Waugh, *The Ordeal of Gilbert Pinfold*, London: Chapman & Hall, 1957, p. 111.
21 A Portuguese version of the document is extant and is in Amsterdam's municipal archives. An English translation, referring to Spinoza's 'abominable heresies', is quoted by Steven Nadler, 'Why Spinoza was excommunicated', *Humanities*, September/October 2013, 34 (5), available at: https://www.neh.gov/article/why-spinoza-was-excommunicated
22 I should be candid about how this was. In P. G. Wodehouse's novel *Joy in the Morning*, Bertie Wooster offers to buy Jeeves any gift he desires and receives the answer: 'Well, sir, there has recently been published a new and authoritatively annotated edition of the works of the philosopher Spinoza. Since you are so generous, I would appreciate that very much.' Bertie wonders if his manservant has got the name right and asks if Spinoza is the Book Society's Choice of the Month. Jeeves's

instincts led me in my teens to seek out just such an edition and I found it beyond me so put it away.

23 Benedict de Spinoza, from *Theological-Political Treatise*, included in *A Spinoza Reader: The Ethics and Other Works*, edited and translated by Edwin Curley, Princeton, NJ: Princeton University Press, 1994, p. 28.

24 Quoted by Peter Watson in a small masterpiece of historical synopsis, *Ideas: A History from Fire to Freud*, London: Weidenfeld & Nicolson, 2005, p. 685. Watson summarises an influential thesis by the historian Jonathan Israel concerning the Enlightenment (p. 687): 'It was Spinoza, [Israel] says, who finally replaced theology with philosophy as the major way to understand our predicament, and as the underpinning rationale of politics; it was Spinoza who dispensed with the devil and magic; it was Spinoza who showed that knowledge is democratic – that there can be no special-interest groups (such as priests, lawyers or doctors) where knowledge is concerned; it was Spinoza who more than anyone persuaded us that man is a *natural* creature, with a rational place in the animal kingdom.' There's vigorous scholarly debate about Israel's thesis of Spinoza's centrality to the Enlightenment, but this passage seems to me unimpeachable as a description of the liberal approach to knowledge.

25 George W. Brown, 'Social roles, context and evolution in the origins of depression', *Journal of Health and Social Behavior*, 43 (3), September 2002, p. 270.

26 Ernst Mayr, *What Evolution Is*, London: Weidenfeld & Nicolson, 2002, pp. 281–2.

27 I owe this observation to a tongue-in-cheek article by S. J. Olshansky, B. A. Carnes and R. N. Butler, 'If humans were built to last', *Scientific American*, 284 (3), 2001, pp. 50–55. The authors further propose that humans would be better off with added ribs, to prevent hernias and other problems by holding organs in place more effectively; thicker bones, to prevent them

from breaking; extra muscles and fat; larger hamstrings and tendons; and knees that are able to bend backwards.

28 Mitchell G. Newberry, Christopher A. Ahern, Robin Clark and Joshua B. Plotkin, 'Detecting evolutionary forces in language change', *Nature*, 551, November 2017, pp. 223–6.

29 Cited by Peter Trudgill, *Dialect Matters: Respecting Vernacular Language*, Cambridge: Cambridge University Press, 2016, p. 1.

30 One small caveat is in order here. Noam Chomsky has long claimed that language is a perfect system. By this he means not a perfect system for communication, but for connecting representations of sound and meaning within the mind of an individual. This intriguing if curious idea is far from being a universal view among linguists, and it doesn't affect my wider point.

31 Jerry A. Coyne, *Why Evolution is True*, Oxford: Oxford University Press, 2009, p. 86.

32 Bullmore, op cit., p. 159.

33 British Neuroscience Association, 'Inflamed depression: An interview with Professor Ed Bullmore', 14 March 2019, available at: https://www.bna.org.uk/mediacentre/news/inflamed-depression/

34 See, for example, the argument of J. M. Eagles, 'Seasonal affective disorder: A vestigial evolutionary advantage?', *Medical Hypotheses*, 63 (5), 2004, p. 767: 'The typical symptoms of recurrent winter depression include lowered mood, lethargy, hypersomnia, social withdrawal, decreased libido, increased appetite and weight gain. Mild hypomania often occurs in spring and summer. It is argued that this pattern of attenuated hibernation constituted an adaptive evolutionary mechanism which enhanced the likelihood of reproductive success, most notably for females, among populations living at temperate latitudes. Women were more likely to become pregnant in the summer and thus to give birth at a time of year when their

babies had a higher chance of survival.' The question mark in the title is telling. Neither the author nor anyone else has any real idea.

Chapter 4: How We Misunderstand Depression

1 Sally Brampton obituary, *The Times*, 12 May 2016 – from which I've taken the quote from her book.
2 Stephen Tindale obituary, *The Times*, 7 July 2017.
3 The programme was broadcast live on ITV on 3 May 2019.
4 Frank Furedi, 'Get off that couch', the *Guardian*, 9 October 2003 – an extract from his book *Therapy Culture*, London: Routledge, 2003.
5 Joanna Williams, 'Stop scaring kids stiff about coronavirus', *Spiked*, 15 April 2020, available at: https://www.spiked-online.com/2020/04/15/stop-scaring-kids-stiff-about-coronavirus/
6 Furedi, op cit.
7 The general position I've outlined is far from unique to the conservative wing of politics and it has some academic proponents. For example, Simon Gottschalk, an American sociologist who identifies as a Democrat, claims that 'some cultural practices today routinely infantilize large swaths of the population' and 'our social institutions and technological devices seem to erode hallmarks of maturity: patience, empathy, solidarity, humility and commitment to a project greater than oneself'. Among the things he complains about are shorter sentences in novels, time spent on smartphones and – which is my point – 'the rise of a "therapy culture"', for which his source is Furedi. He offers no evidence that these things are connected, or that they exist at all, let alone that they have the deleterious effects he imagines. See Simon Gottschalk, 'The infantilization of western culture', *The Conversation*, 1 August 2018, available at: https://theconversation.com/the-infantilization-of-western-culture-99556

8 Janet Street-Porter, 'Depression? It's just the new trendy illness!', the *Daily Mail*, 5 August 2010.

9 Greg Hurst, interview with Bill Oddie, 'Celebrities are making mental illness fashionable', *The Times*, 11 November 2013.

10 Grace Campbell, 'Alastair Campbell's daughter, Grace, shares lessons from living with a parent suffering from depression', the *Sunday Times*, 16 June 2019.

11 J. Stone et al., 'What should we say to patients with symptoms unexplained by disease? The "number needed to offend"', *BMJ*, 21 December 2002, pp. 1449–50.

12 Miguel de Unamuno, translated by J. E. Crawford Flitch, *Tragic Sense of Life*, New York: Dover Publications, 1954, p. 17.

13 Theodore Dalrymple, review of Clark and Layard, op cit., *The Times*, 19 July 2014.

14 The return of measles to Costa Rica was reported in February 2019, after the country had been free of the disease for five years. The probable reason was the visit on holiday of a young French boy, with his family, who had not received the MMR vaccine. According to the WHO, vaccination resulted in an 80 per cent decline in deaths from measles globally between 2000 and 2017, yet there are still 110,000 such deaths – mainly of children under five – each year. Figures taken from WHO factsheet, *Measles*, 9 May 2019, available at: https://www.who.int/news-room/fact-sheets/detail/measles

15 Allan Horwitz and Jerome Wakefield, *The Loss of Sadness: How Psychiatry Transformed Normal Sorrow into Depressive Disorder*, New York: Oxford University Press, 2007.

16 Interview by Jonah Lehrer, 'Is There Really an Epidemic of Depression?', *Scientific American*, 4 December 2008, available at: https://www.scientificamerican.com/article/really-an-epidemic-of-depression/

17 Steven Pinker, *Enlightenment Now: The Case for Reason, Science, Humanism and Progress*, London: Penguin, 2018, p. 281.

18 One example is a statement by 442 therapists, counsellors

and academics ahead of the 2015 general election, issued in the form not of a scholarly paper but of a round-robin letter to the *Guardian*. It argued that 'the wider reality of a society thrown completely off balance by the emotional toxicity of neoliberal thinking is affecting Britain in profound ways, the distressing effects of which are often most visible in the therapist's consulting room. This letter sounds the starting-bell for a broadly based campaign of organisations and professionals against the damage that neoliberalism is doing to the nation's mental health.' There's a strong case (which, as an economics writer, I've often argued) that unnecessarily rapid fiscal consolidation in the years after the financial crash starved social services of resources and intensified in-work poverty, but that's not what the critics are arguing. Rather they're stating that mental illness is caused by 'neoliberal thinking'. The field of mental health is not on sufficiently empirical a footing to be able to generate political conclusions like this, or at least to do so legitimately. Many of the letter's signatories are practitioners of therapies, such as Jungian analysis, that have little or no scientific support. The causes of mental disorder are enough of a conundrum already; hypothesising a link with 'neoliberalism', which the letter fails to define, doesn't explain anything. See letter, 'Austerity and a malign benefits regime are profoundly damaging mental health', the *Guardian*, 17 April 2015.

19 See Laurie Penny, 'Don't give in: An angry population is hard to govern; a depressed population is easy', the *New Statesman*, 9 May 2015, for an example of this genre. Penny writes: 'There's a reason depression and its precarious cousin, anxiety, are the dominant political modes of late capitalism. This is how you're supposed to feel.' She cites no sources who have expressed what you're 'supposed' to feel. A more academic but no better-evidenced form of the same thesis can be found in the writings of the late cultural theorist Mark Fisher, notably *Ghosts of*

My Life: Writings on Depression, Hauntology and Lost Futures, Winchester: Zero Books, 2014.

20 Pinker, op cit., 2018, p. 282.

Chapter 5: Diagnosing Depression

1 Waugh, op cit, p. 111.

2 Lord Byron, 'Jephthah's Daughter', *The Complete Poetical Works*, edited by Jerome J. McGann, 7 vols, vol. 3, Oxford: Clarendon Press, 1981, p. 294.

3 John Cheever, *The Housebreaker of Shady Hill and Other Stories*, London: Victor Gollancz, 1958, p. 29.

4 The classic study of this phenomenon is Richard Hofstadter's *The Paranoid Style in American Politics and Other Essays*, New York: Alfred A. Knopf, 1965. Hofstadter identifies the paranoid style as a conspiratorial world view that he traces from the belief of New England clergy in the late eighteenth century that the Bavarian Illuminati were the secret force behind the Jeffersonians, through various historical guises up to McCarthyism and the contemporary radical Right. The extent to which conspiracy theories reflect everyday cognitions has become a subject for study in social psychology in its own right.

5 Otto Neurath, 'Protocol Statements', 1932, cited in Peter J. Mehl, 'Kierkegaard and the relativist challenge to practical philosophy', *The Journal of Religious Ethics*, 14 (2), Fall 1986, p. 274.

Chapter 6: Medical Treatment

1 Nathaniel Lee, *Cesare Borgia*, Act 5, Scene 1, quoted in Jane E. Kromm, 'Hogarth's madmen', *Journal of the Warburg and Courtauld Institutes*, 48, 1985, p. 238.

2 Evelyn A. Woods and Eric T. Carlson, 'The psychiatry of Philippe Pinel', *Bulletin of the History of Medicine*, 35 (1), January/February 1961, p. 18.

3 This history is recounted in Anne Digby, 'Changes in the asylum: The case of York, 1777–1815', *The Economic History Review*, New Series, 36 (2), May 1983, pp. 218–39.
4 This tragic life is recounted in Kate Larson Clifford, *Rosemary: The Hidden Kennedy Daughter*, New York: Mariner Books, 2016.
5 John F. Fulton, 'The physiological basis of psychosurgery', *Proceedings of the American Philosophical Society*, 95 (5), 17 October 1951, p. 539.
6 Both are quoted in Nicola Davis and Pamela Duncan, 'Electroconvulsive therapy on the rise again in England', the *Guardian*, 17 April 2017.
7 Mariam Alexander, 'As a psychiatrist, if I had severe depression I'd choose ECT', the *Guardian*, 22 July 2019.
8 Information taken from Mind, the mental health charity, at: https://www.mind.org.uk/information-support/drugs-and-treatments/electroconvulsive-therapy-ect/#.XQkR3rzYrnE
9 I owe this explanation and some of the following technical descriptions of antidepressants to Blaise A. Aguirre, *Depression* (Biographies of Disease series), Westport, CT: Greenwood Press, 2008, Chapter 4.
10 Lucas Richert, *Break on Through: Radical Psychiatry and the American Counterculture*, Cambridge, MA: MIT Press, 2019, p. 147.
11 Studies of microdosing on psychedelic drugs are few in number and predominantly anecdotal, but they generally agree that the practice originated in Silicon Valley among millennials and then spread.
12 Peter D. Kramer, *Listening to Prozac*, London: Penguin, 1993, especially the Appendix, 'Violence', pp. 301–14.
13 Peter D. Kramer, 'Prozac: Better than well', letter in *The Lancet*, January 2016, available at: https://www.thelancet.com/journals/lanpsy/article/PIIS2215-0366(15)00552-0/fulltext
14 A good example of this is a collected volume, Carl Elliott

and Tod Chambers (eds), *Prozac as a Way of Life* (Studies in Social Medicine), Chapel Hill, NC: University of North Carolina, 2004, p. 7. According to the editors: 'Every culture has its own socially prescribed psychoactive substances, from peyote, kava, and betel nuts to alcohol, caffeine, and nicotine. But with the SSRIs, the gate to the drug is guarded by doctors, and the passport for access is the diagnosis of a mental disorder. Unlike alcohol, which is dispensed in bars and liquor stores, or caffeine, which is dispensed at Starbucks and Unitarian churches, SSRIs are dispensed at doctor's offices and pharmacies. It is the social place occupied by the SSRIs that has produced the ambivalence that many of us feel about their popularity. Unlike bartenders and espresso baristas, doctors have not generally thought of their job as making well people feel better than well. But that might change.'

15 Irving Kirsch, *The Emperor's New Drugs: Exploding the Antidepressant Myth*, London: Bodley Head, 2009, pp. 5–6.

16 'Position statement of the European Psychiatric Association (EPA) on the value of antidepressants in the treatment of unipolar depression', *European Psychiatry*, 27 (2), February 2012, pp. 114–128.

17 Fran Lowry, 'APA Blasts *60 Minutes* Program on Antidepressants', *Medscape Medical News*, 20 March 2012, available at: https://www.medscape.com/viewarticle/760569

18 Johann Hari, 'Is everything you think you know about depression wrong?', the *Observer*, 7 January 2018. It's an extract from Hari's book, which I haven't read.

19 National Institute for Health and Care Excellence, available at: https://www.nice.org.uk/guidance/cg90/ifp/chapter/Treatments-for-moderate-or-severe-depression

20 NHS, 'Treatment: Clinical depression', available at: https://www.nhs.uk/conditions/clinical-depression/treatment/

21 Dean Burnett, 'Is everything Johann Hari knows about depression wrong?', the *Guardian*, 8 January 2018, available

at: https://www.theguardian.com/society/2018/jan/07/is-everything-you-think-you-know-about-depression-wrong-johann-hari-lost-connections?CMP=fb_gu

22 I argued this point in a column, 'Let's hope there's not a lot of Conan Doyle in Sherlock', *The Times*, 31 December 2015, and have rarely had so much (generally negative) reader correspondence. Holmes enthusiasts, who enjoy dressing up in period costume and maintain a museum at the invented address of 221B Baker Street, are a devoted band.

23 A. Cipriani, T. A. Furukawa, G. Salanti et al., 'Comparative efficacy and acceptability of 21 antidepressant drugs for the acute treatment of adults with major depressive disorder: A systematic review and network meta-analysis', *The Lancet*, 21 February 2018.

24 News release, 'Antidepressants are more effective than placebo at treating acute depression', Department of Psychiatry, Medical Services Division, University of Oxford, 22 February 2018, available at: https://www.psych.ox.ac.uk/news/all-antidepressants-are-more-effective-than-placebo-at-treating-acute-depression-in-adults-concludes-study

25 NHS website, 'Big new study confirms antidepressants work better than placebo', 22 February 2018, available at: https://www.nhs.uk/news/medication/big-new-study-confirms-antidepressants-work-better-placebo/

26 On this study and doctors' response to it see Thomas Sexton, 'Paxil brings suicidal thoughts', *Psychology Today*, 14 July 2003 (reviewed on 9 June 2016), available at: https://www.psychologytoday.com/us/articles/200307/paxil-brings-suicidal-thoughts

27 My comments in the following paragraphs about the effects of antidepressants draw on Donald F. Klein MD and Paul H. Wender MD, *Understanding Depression: A Complete Guide to Its Diagnosis and Treatment*, New York: Oxford University Press, 2005 (2nd edn), especially Chapter 5. Whereas most

clinically informed opinion (as I've cited above) recommends a combination of medication and psychological treatment for cases of moderate to severe depression, the authors recommend antidepressants as the first recourse, on the grounds that 'inappropriate psychotherapy not only can delay the relief from painful symptoms (with consequent bad effects on the patient's personal, vocational and social life) but also can make the patient feel even more helpless when the therapy fails to produce the anticipated improvement'. This is indeed what happened to me, and medication was the first step I took towards recovery; but psychological therapy was the more important factor over the longer term.

28 Quoted by Matt Seaton, 'The wit and wisdom of Tony Benn', the *Guardian*, 31 May 2007.

29 For example, see Chloe Lambert, 'Critics claim antidepressants are being handed out like sweets. Now our shocking experiment uncovers . . . The proof doctors are doling out happy pills to anyone who asks', the *Daily Mail*, 30 September 2013. There is no 'proof' cited in this article, and the purported experiment consisted of sending three women to their GPs to lie about their symptoms. The author arbitrarily determines that this subterfuge, to which the doctors responded promptly and properly, provides 'an alarming snapshot of what appears to be a national trend: our growing appetite for the pills'.

30 Ian C. Reid, 'Are antidepressants overprescribed? No', *BMJ*, 346, 11 May 2013, p. 16. The opposite view is advanced in the same issue by Des Spence, a Glasgow GP who is part of an advocacy organisation that seeks 'to limit the influence of Big Pharma over drug promotion and education'. Both articles are available here, and they exemplify the difference in approach between a scholar's concern for evidence and a campaigner's zeal: https://www.bmj.com/bmj/section-pdf/187887?path=/bmj/346/7907/Head_to_Head.full.pdf

31 Christopher Hitchens, *Hitch-22: A Memoir*, London: Atlantic Books, 2010, p. 342.
32 Aldous Huxley, *Brave New World*, New York: Harper & Brothers, 1932, p. 96.

Chapter 7: Psychological Treatment

1 Frederick Crews, *Follies of the Wise: Dissenting Essays*, New York: Shoemaker & Hoard, 2006, p. 70.
2 Mary Cregan, *The Scar: A Personal History of Depression and Recovery*, New York: W. W. Norton & Company, 2019, p. 200.
3 'Training to become a counsellor or psychotherapist', British Association for Counselling and Psychotherapy, available at: https://www.bacp.co.uk/careers/careers-in-counselling/training/
4 May Bulman, 'Mentally ill "exploited" by unaccredited online counselling', the *Independent*, 13 January 2018.
5 National Institute for Health and Care Excellence, 'Depression in adults: Recognition and management', October 2009, last updated April 2018, available at: https://www.nice.org.uk/guidance/cg90/ifp/chapter/treatments-for-mild-to-moderate-depression
6 There is less psychodynamic therapy in the National Health Service than there once was, but it's practised in these fields and has been influenced by what's known as Object Relations Theory. This approach, which is associated especially with the Austrian-British psychoanalyst Melanie Klein, emphasises the role of human contact and the need to form relationships in the development of personality. It can be valuable and even crucial in treating some of the very unwell people whom psychiatrists see in the NHS who may not have had a single positive relationship in their lives.
7 My family edition is George Eliot, *The Mill on the Floss*, Edinburgh: William Blackwood & Sons, 1907, pp. 683–4, originally published in 1860. By referring to a 'natural

priesthood', Eliot is clearly gesturing to the notion of the secular 'religion of humanity' devised by Auguste Comte, the French positivist philosopher. Comte was a great social thinker but some of his ideas were very strange, including this, and Eliot wrote scathingly and wisely in the 1870s of his advocacy of a new priesthood: 'Doctrine, no matter of what sort, is liable to putrefy when kept in close chambers to be dispensed according to the will of men authorized to hold the keys.' Her very moving authorial aside in the passage I've quoted is best thought of as a simple expression of people brought together by a common humanity. More on Eliot's views of this movement (including the quotation I've just cited) is set out in Rosemary D. Ashton, 'The intellectual "medium" of *Middlemarch*', *The Review of English Studies*, 30 (118), May 1979, pp. 154–68.

8 Rhodri Marsden, 'Stopping therapy: We have ways of making you talk', the *Independent*, 17 June 2014.

9 Beck, op cit., p. 9.

10 Cited by Daniel B. Smith, 'The doctor is "in": At 88, Aaron Beck is now revered for an approach to psychotherapy that pushed Freudian analysis aside', *The American Scholar*, 78 (4), Autumn 2009, p. 21. I've drawn on this article in summarising the techniques that Beck devised, and also on Richard Layard and David M. Clark, op. cit., especially pp. 8–15 and 120–27, as well as Beck's own writing.

11 Moreover, many clinical psychologists are in the NHS and work in multi-disciplinary teams. In that context a patient will come to the team by a variety of routes, including referral by a GP, and maybe a variety of professional groups who work with a given patient.

12 Layard and Clark, op cit., p. 9.

13 Robert Burton, *The Anatomy of Melancholy*, 5 vols, vol. 1, edited by Thomas C. Faulkner, Nicolas K. Kiessling and Rhonda L. Blair, Oxford: Clarendon Press, 1989, pp. lxx-lxxi (first published 1621).

14 Beck, op cit., p. 7.
15 Cited by Smith, op cit., p. 22.
16 Scott O. Lilienfeld, Steven Jay Lynn, John Ruscio and Barry L. Beyerstein, *50 Great Myths of Popular Psychology: Shattering Widespread Misconceptions about Human Behavior*, Chichester: John Wiley & Sons, 2010, p. 238.
17 From *The Collected Poems of Stephen Crane*, 1930, quoted in Martin Gardner, *The Whys of a Philosophical Scrivener*, Oxford: Oxford University Press, 1983, p. 360.
18 Kevin J. Flannelly and Kathleen Galek, 'Religion, evolution, and mental health: Attachment theory and ETAS theory', *Journal of Religion and Health*, 49 (3), September 2010, p. 339.
19 Agency for Healthcare Research and Quality, *Meditation Programs for Psychological Stress and Well-Being*, prepared by Johns Hopkins University Evidence-based Practice Center, Baltimore, MD, 2014, p. viii, available at: https://effectivehealthcare.ahrq.gov/sites/default/files/pdf/meditation_research.pdf
20 Quoted in Alvin Powell, 'When science meets mindfulness', *The Harvard Gazette*, 9 April 2018, available at: https://news.harvard.edu/gazette/story/2018/04/harvard-researchers-study-how-mindfulness-may-change-the-brain-in-depressed-patients/

Chapter 8: Living With Depression

1 Beck, op cit., p. 49.
2 Ibid.
3 I've taken these figures from a survey by the Office for National Statistics, *Families and Households: 2018*, released on 7 August 2019, and available at: https://www.ons.gov.uk/peoplepopulationandcommunity/birthsdeathsandmarriages/families/bulletins/familiesandhouseholds/2018
4 *The Times* once had an etiquette expert, John Morgan, who dispensed advice weekly in response to readers' queries. In one

column he cited a reader who wished to find out about flirting – what the concept meant, how it was done, and whether there was anything he could read on the subject. Morgan deftly and sensitively said that he was sadly unable to offer any useful advice. I was perhaps equivalent in hoping to find out about the qualities of the human heart through reading about them, but I did come across an excellent and unjustly little-known philosophical exploration of the themes of sexual and romantic attraction by Anne Kelleher, *Sex Within Reason*, Jonathan Cape, 1987, whose central point about the character of love I recount in this paragraph and in Chapter 2.

5 Dante Alighieri, *The Vision; or Hell, Purgatory, and Paradise*, Canto XXI, translated by Rev. H.F. Cary, London: Frederick Warne & Co., 1814, p. 281.

Chapter 9: Depression and Art

1 Boswell, op cit., p. 102.
2 Ibid., p. 1026.
3 Ibid., p. 169.
4 Ibid., p. 316.
5 Quoted by D. Nichol Smith, 'Samuel Johnson's poems', *The Review of English Studies*, 19 (73), January 1943, p. 47.
6 Samuel Johnson, 'Death', *The Idler*, no. 41, 27 January 1759.
7 James Boswell, 'An Account of My Last Interview with David Hume, Esq., Partly recorded in my Journal, partly enlarged from my memory', 3 March 1777, reproduced by the National Library of Scotland and available at: https://digital.nls.uk/scotlandspages/timeline/17762.html
8 Michael Ignatieff, *The Needs of Strangers*, London: Chatto & Windus, 1984, p. 101.
9 Boswell, op cit., p. 511.
10 Ibid., p. 633.
11 Quoted by Boswell, ibid., p. 385.

12 Scull, op cit., p. 94
13 Burton, op cit., p. 69.
14 Ibid., p. 243.
15 Ibid., p. 65.
16 The spread of this curious view is neatly summarised in Clark Lawlor, *From Melancholia to Prozac: A History of Depression*, Oxford: Oxford University Press, Chapter 2.
17 *As You Like It*, Act 4, Scene 1.
18 John Milton, 'Il Penseroso', in *The Complete Works of John Milton*, 11 vols, vol. 3, edited by Barbara Kiefer Lewalkski and Estelle Haan, Oxford: Oxford University Press, 2012, p. 32.
19 There are many editions of Keats. The poem is in a transatlantic anthology I especially value, W. H. Auden and Norman Holmes Pearson (eds), *The Portable Romantic Poets: Blake to Poe*, New York: Viking Penguin Inc., 1950, pp. 390–91.
20 Robert Cummings, 'Keats's melancholy in the temple of delight', *Keats-Shelley Journal*, 36, 1987, p. 52.
21 Scull, op cit., p. 94.
22 Quoted in Judith Farr, *The Passion of Emily Dickinson*, Cambridge, MA: Harvard University Press, 1992, p. 128.
23 Quoted in ibid., p. 5.
24 Mary Ann Wells, 'Was Emily Dickinson psychotic?', *American Imago*, 19 (4), Winter 1962, p. 313. I've drawn on this useful essay for other biographical details too. I know nothing about Wells apart from her books. She wrote a handful of excellent mystery novels and also a biography of Thomas Wentworth Higginson, the long-standing correspondent, mentor and posthumous editor of Emily Dickinson.
25 All references in what follows come from Emily Dickinson, *The Complete Poems*, edited by Thomas H. Johnson, London: Faber & Faber, 1975. The numbers in brackets indicate the number of the poem in this standard edition.
26 *Macbeth*, Act 5, Scene 3.
27 F. R. Leavis, *Nor Shall My Sword: Discourses on Pluralism*,

Compassion and Social Hope, London: Chatto & Windus, 1972, p. 98.

28 Browning, op cit., pp. 645–6.

29 Steven Pinker, *The Language Instinct*, London: Penguin, 1994, p. 82.

30 I owe these observations on Ruskin, and quotes from him about Turner, to Michael Wheeler, *Ruskin's God*, Cambridge: Cambridge University Press, 1999, p. 47.

31 I set out the case for the prosecution some years ago when Chesterton was being proposed (unsuccessfully, as it turned out) for canonisation, in 'G. K. Chesterton: A writer unfit to be a saint', *Jewish Chronicle*, 29 August 2013, available at: https://www.thejc.com/comment/columnists/g-k-chesterton-a-writer-unfit-to-be-a-saint-1.48179

32 G. K. Chesterton, 'The Glory of Grey', in *Alarms and Discursions*, London: Methuen, 1910, p. 120.

Epilogue: The End of Depression

1 Quoted (in an apt synthesis) by Scull, op cit., p. 94.

Index